1 Forces

Forces can ...

▶ change the shape or size of an object.

▼ change the speed of an object.

The force of the engine makes the car go faster

The pole bends

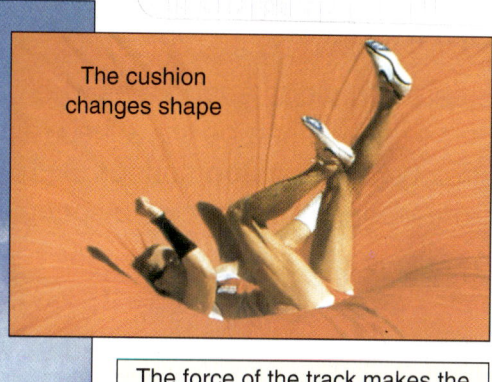
The cushion changes shape

The force of the track makes the cars go round the corkscrew

▶ change the direction of movement of an object.

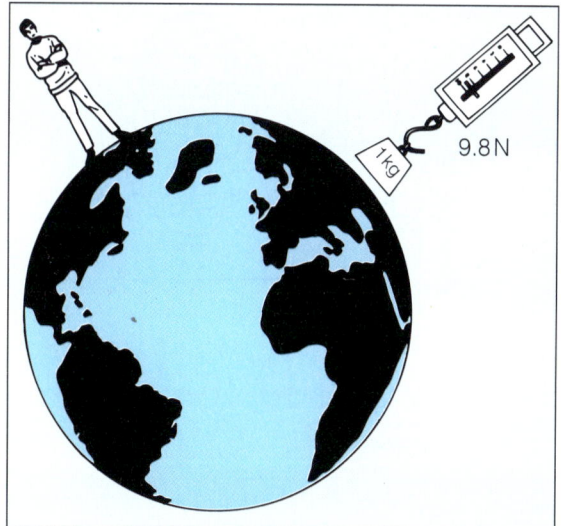

▲ Weight is a force. It is the force of gravity. Everything is pulled towards the centre of the Earth. The bigger an object the greater the force of attraction. It is everywhere and we cannot turn it off.

▶ A spring balance is used to measure force. Force is measured in units called **newtons**. 1 kg has a weight of 10 N (more exactly 9.8 N).

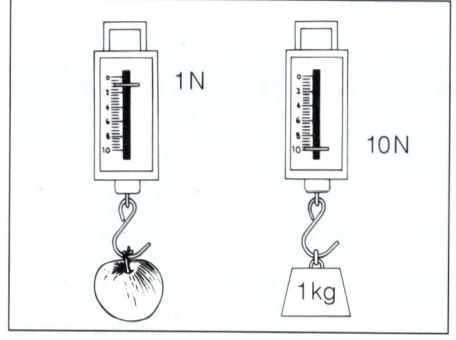

▲ Forces always act in pairs. This ice skater has two forces acting on her. The man pushes up and her weight pushes down.

1

2 Forces and elasticity

Springs

Steel changes its shape a bit when loaded. We use this effect in springs. Some springs can be squashed (a **compression** spring), others can be stretched (**extension** spring). Both types behave in the same way. Let's increase the load force on a spring and measure the extension produced.

Q1 Copy and complete:
"I predict that if I double the load, the extension will _____"

Q2 Copy this table.

Load (N)	Pointer reading (cm)	Extension (cm)
0		
1		
2		

Apparatus

- ☐ 2 clamps and stands
- ☐ pin ☐ extension spring
- ☐ set of slotted weights and hanger (masses will do)
- ☐ metre rule ☐ Plasticine
- ☐ thin copper wire
- ☐ low powered microscope or magnifying glass
- ☐ eye protection

Eye protection must be worn when stretching the spring and wire in case it breaks and whips.

A Fix the spring to the clamp. Put the hanger on the lower end of the spring. ▼

B Use plasticine to fix the pin on to the hanger. Put the rule in the other clamp. The '0' end should be at the top. ▼

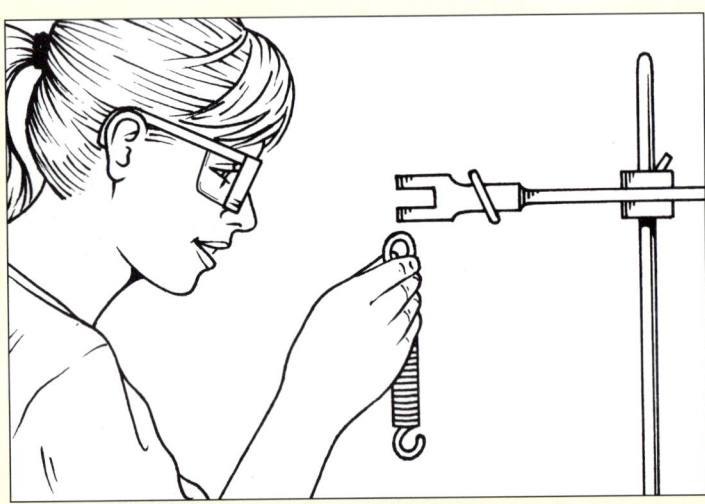

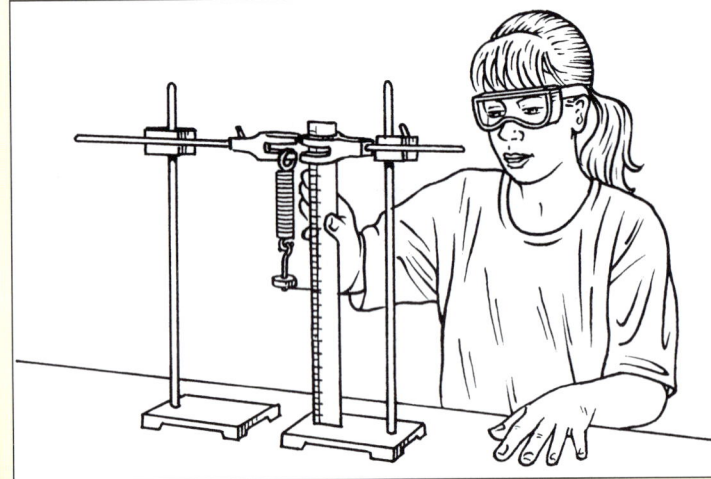

C Adjust the rule so that the pin points to a number. This is your starting point. Record the scale reading in your table. ▼

D Add another weight. Note the pointer reading. Repeat **D** four more times. Complete your table each time. ▼

E Unload the spring to check that it returns to its original reading. Draw a graph of extension against load. Use these axes. ▶

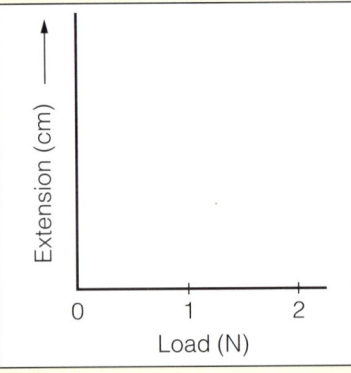

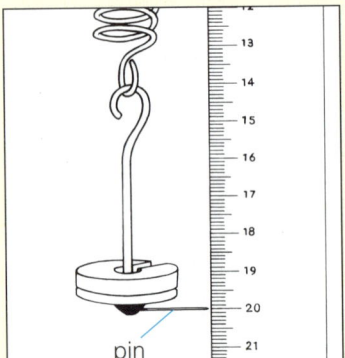

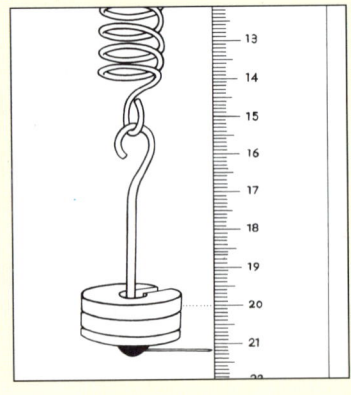

Q3 What was the shape of your graph?

Q4 Was your prediction correct?

Q5 In **E** did your spring return to its original length?

2 Forces and elasticity

Over the limit

If your graph was a straight line, then when you doubled the load you doubled the extension. When the graph is a straight line like this we say: 'The extension is **directly proportional** to the load.' This is also known as Hooke's law. It applies to car springs, bed springs and even to bridges.

The question is: Is there a limit to the load you can add to the spring?

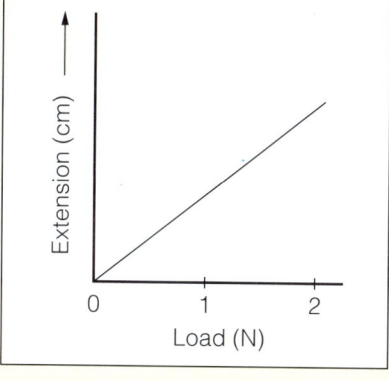

Q1 Predict how the spring will behave when you keep loading it.

Let's load a spring to find out what happens. Springs are expensive so your teacher may demonstrate this for you.

F Your teacher will repeat **A** to **D**, adding weights until there is a sudden change to the spring.

G Unload the spring. Note if the pointer reading returns to its original position.

At breaking point

Now let's stretch a wire until it breaks.

 Wear eye protection.

 Take care of toes and floor.

H Replace the spring with thin copper wire. Make another copy of the table in **Q2** on page 2. Repeat **A** to **D**. Continue loading the wire until it breaks. ▼

I Examine the broken ends of the wire with a low powered microscope. ▼

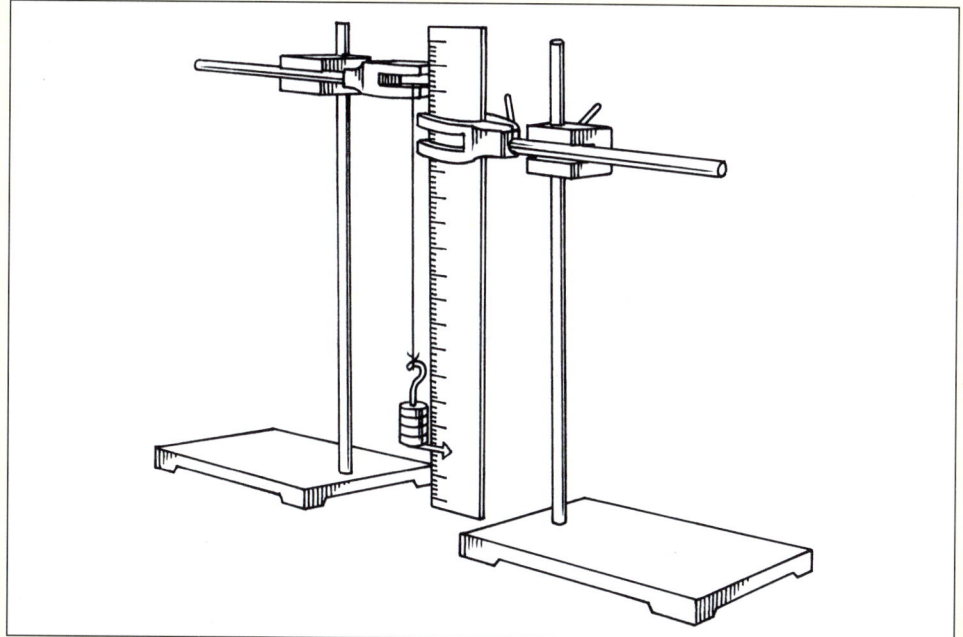

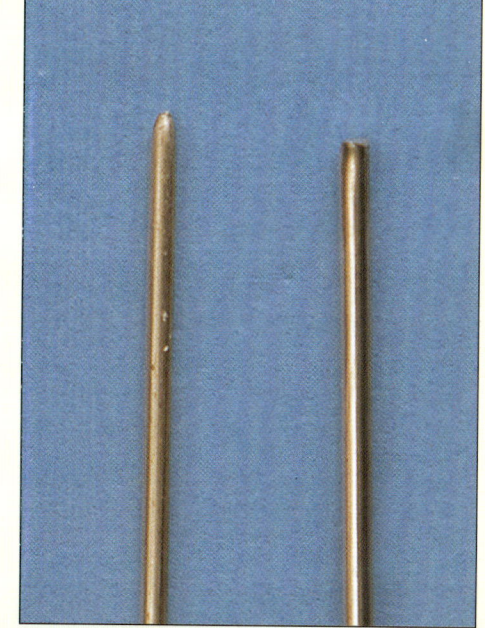

Q2 Draw a graph of extension against load for the spring in **F** and **G**.

Q3 Was your prediction for **Q1** correct?

Q4 a What suddenly happened to the spring in **F**?
b What did you notice when the spring was unloaded in **G**?

Q5 What was the breaking point of your wire?

Q6 Sketch the appearance of the broken end of the wire.

3

Forces and elasticity

Tension force

Forces which pull and stretch a material are called **tension** forces. The amount by which the material stretches is called the extension.

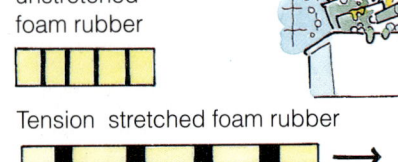

Compression force

Forces which squash or push are called compression forces.

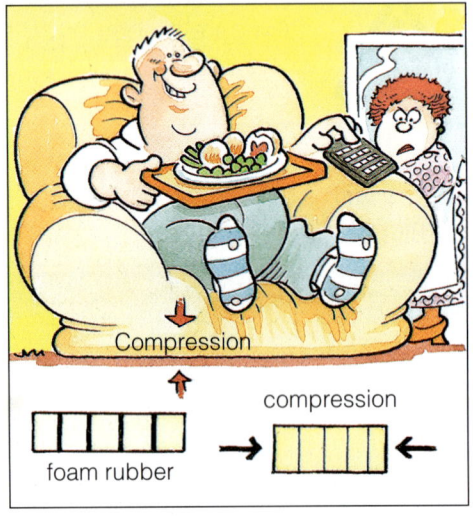

Bending force

With a bending force, the top surface is in tension, the lower surface is in compression.

Shear force

When we wring out a wet towel we apply a **shear** or twisting force.

Elasticity

An elastic material is one which returns to its original shape once the **deforming** force is removed.

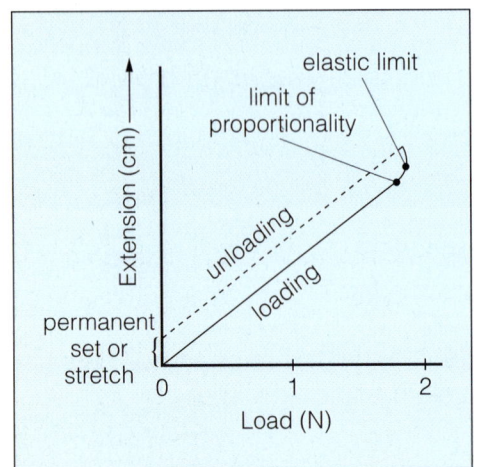

Hooke's law applies to elastic materials. Once the limit of proportionality is exceeded, the law no longer applies. When the elastic limit is exceeded the material has a permanent stretch or set.

Q1 What type of force (tension, compression, etc.) is experienced by:
a a ball struck by a bat
b a screwdriver
c a nail when hit by a hammer
d a nail when pulled out of a piece of wood
e the shaft of a golf club hitting the ball?

3 Turning forces

Balancing forces

▶ A force which acts some distance from a pivot is a **turning force**. We can find how strong a turning force is by using:

turning force = force × distance of force from pivot

A turning force is often called the **moment** of the force.

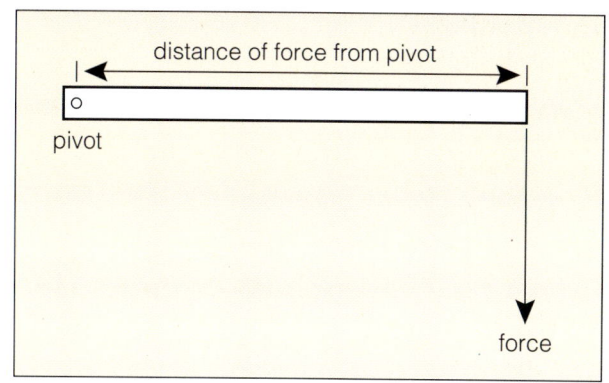

Examples of turning forces

Remember:
moment of a force = force (N) × perpendicular distance (m).

▼ A small force a large distance from the pivot can exert a large turning force. We use this effect in levers.

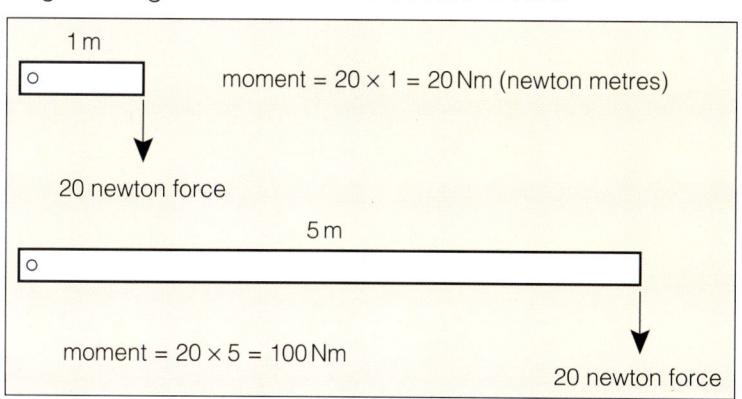

▼ A spanner is a lever. The longer the spanner the easier it is to undo a tight nut.

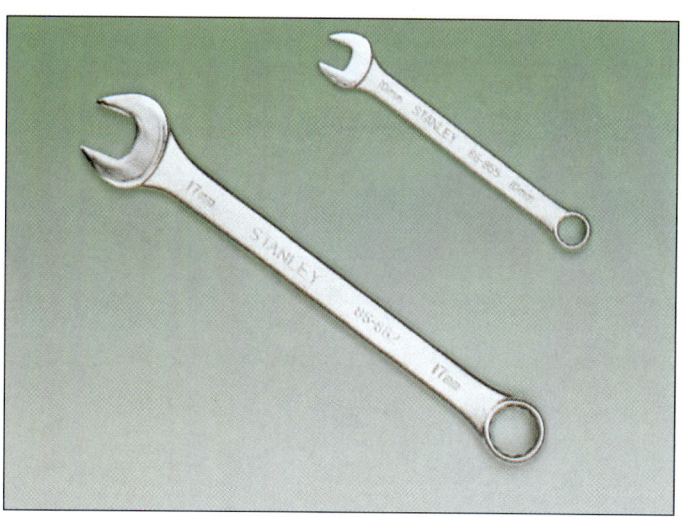

▼ Here are some machines which use levers to make our work easier.

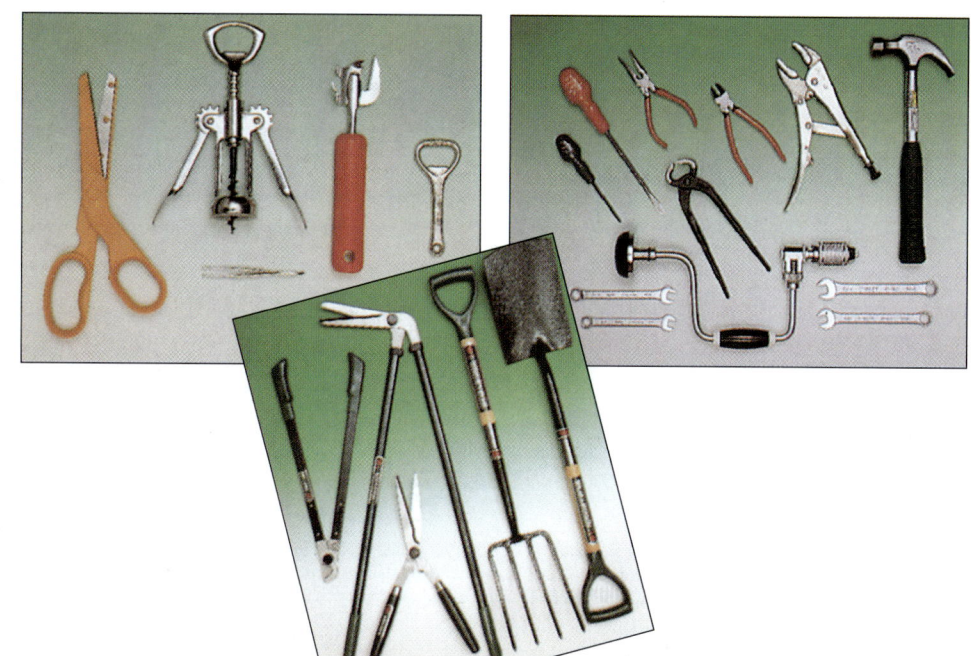

Q1 The cat is on the end of a long lever. It just balances the heavy weight and the mouse.

What would happen if the cat ate some cream?

Q2 What would happen if the cat turned round and walked along the lever to get the mouse?

5

3 Turning forces

Investigating balance

In this experiment let us investigate turning forces (moments) which balance.

Apparatus

☐ metre rule with a hole in the centre ☐ 2 clamps and stands
☐ 3 small loops of fine string
☐ 2 sets of slotted weights and hangers ☐ write-on sticky tape

Q1 Copy this table.

left-hand side			right-hand side		
force (N)	distance of weight from centre hole (cm)	force × distance	force (N)	distance of weight from centre hole (cm)	force × distance
1	10	10	1		
1	20	20	1		
1	30	30	1		
2	10	20	1		
3	10	30	1		

A Fix a piece of write-on sticky tape to a ruler. Mark the centimetre lines and number them as shown. ▼

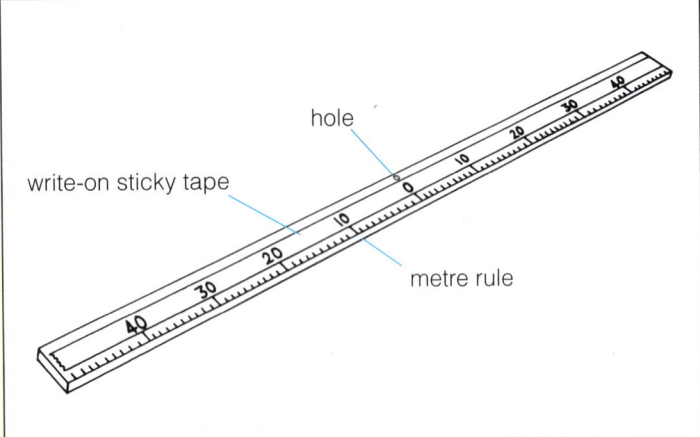

B Thread a small loop of string through the hole and hang the ruler from a clamp. ▼

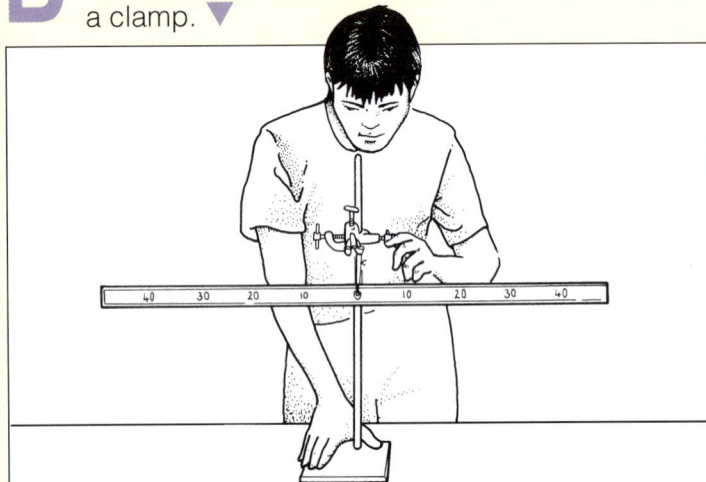

C Use the wide-open jaws of another clamp to stop the ruler from **tilting** too far. ▼

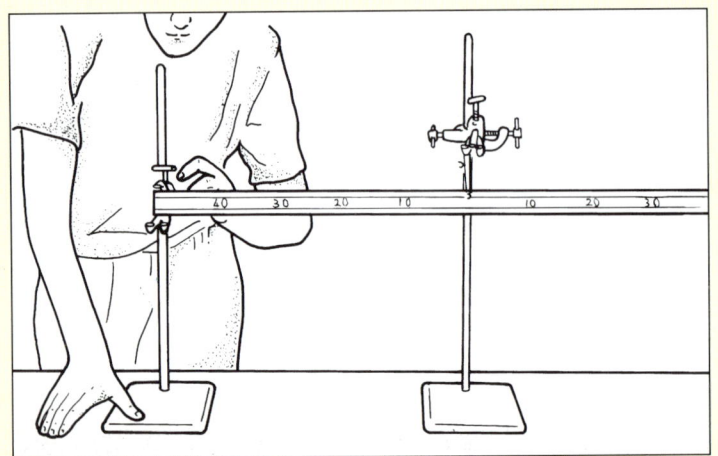

D Put a 1 N hanger on the left-hand side so that it is 10 cm from the hole. Put the other hanger on the right-hand side. Move it until the rule balances. Measure how many centimetres the hanger is from the hole, record your results in the table. ▼

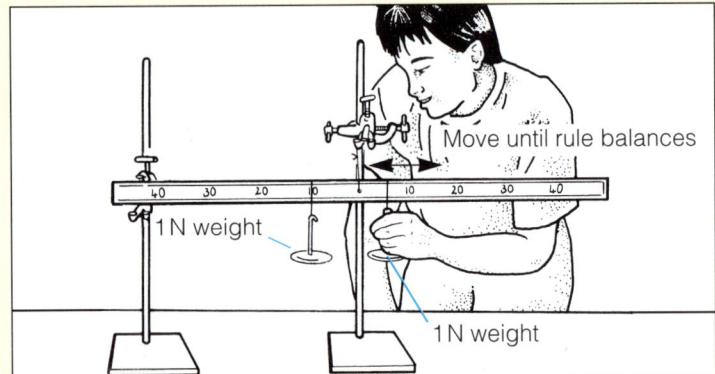

3 Turning forces

E Repeat **D** with the left-hand weight 20 cm from the hole, then 30 cm from the hole. Record your results. ▼

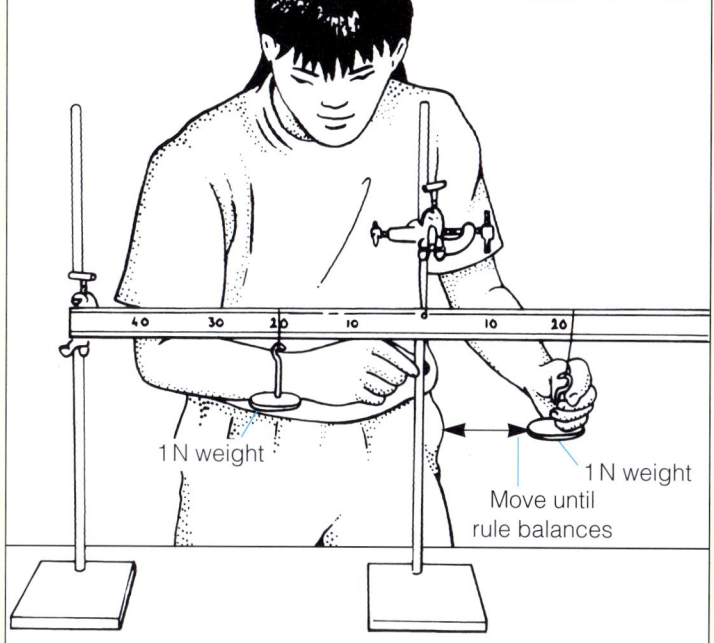

F Change the left-hand weight to 2 N. Hang it 10 cm from the hole.
Move the 1 N weight on the right-hand side until the rule balances. Measure how far the 1 N weight is from the hole. Record your results in the table. ▼

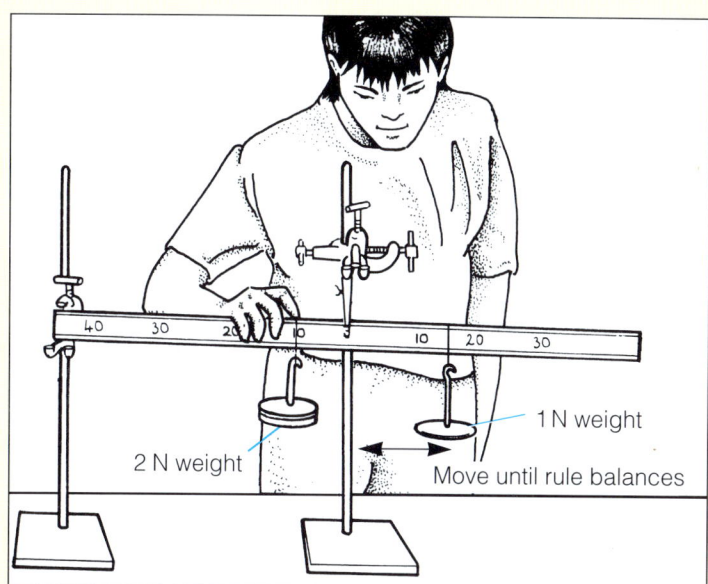

G Repeat **F** using 3 N on the left-hand side. Hang it 10 cm from the hole. ▶

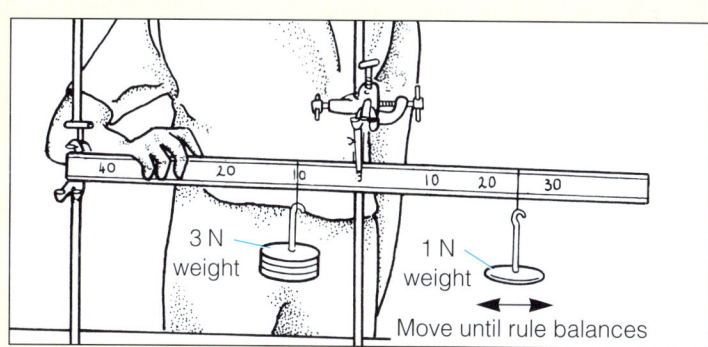

H Look at the third column in the table. The force and the distance have been multiplied (×) together. To complete the last column multiply (×) the force by the distance it is from the hole.

Q2 When a 1 N weight was 10 cm from the hole, how many centimetres from the hole was the other 1 N weight when it balanced.

Q3 When a 2 N weight is 10 cm from the hole, where must a 1 N weight be placed to make it balance?

Q4 Dave's been talking in class again and has not put down all the results of his experiment. Copy and complete the table for him.

Left-hand side			Right-hand side		
Force (N)	Distance of weight from centre hole (cm)	Force × distance	Force (N)	Distance of weight from centre hole (cm)	Force × distance
1	15		1.5		15
1			1	20	
1	25		2.5		
2	15		1		
3	10	30	1		

Extension exercises 1 and 2 can be used now.

4 Balanced forces and velocity

On balance

Let's look at the forces acting on our hero as he delivers some flowers to his girlfriend.

The weight force of the helicopter acts down and is just balanced by the lift from the rotor blades. The helicopter hovers just above the mountain top.

Our hero jumps. His weight force is unbalanced so he **accelerates** (falls faster and faster) as he falls the short distance to Earth. ▲

Skiing down a gentle slope, the downward force due to his weight just balances the **friction** (drag) forces on him. He doesn't speed up or slow down. He travels at **constant** speed. ▲

Driving flat out at 200 km/h. The forward force from the engine just balances the friction (drag) forces. He travels at a constant speed. ▲

Waiting at the door, his weight force down is just balanced by the upward **reaction** force from the ground. ▲

Balanced forces have no effect on the existing motion of a body. If balanced forces act on a stationary body it will not move. ▲

If balanced forces act on a moving body it will continue to move, neither slowing down nor speeding up. ▲

However, an unbalanced force will make a body change its speed or direction of travel. ▲

Q1 What would happen to the helicopter if the lift force is:
a increased
b decreased?

Q2 What would happen to the helicopter as our hero jumps out?

Q3 What effect do balanced forces have on his speed down the ski slope?

Q4 How would the forces change his motion if the slope were:
a steeper
b less steep?

Q5 Which forces act on the car and what effect do they have on its speed?

Q6 Which force(s) would have to change to make the car:
a slow down
b then speed up?

Extension exercise 3 can be used now.

4 Balanced forces and velocity

Speed and velocity

Speed

We can measure how fast something is travelling from the calculation:

$$\text{speed (m/s)} = \frac{\text{distance travelled (m)}}{\text{time taken (s)}}.$$

If the car travels 50 km in one hour.

Its speed is $\frac{50}{1} = 50$ km/h

If the bicycle travels 500 m in 100 seconds.

Its speed is $\frac{500}{100} = 5$ m/s

Velocity

Velocity is speed in a particular direction. We must always say which direction it is moving in as well as how fast. The car has a velocity of 50 km/h due east.

So, when speaking of velocity we must consider the speed in a particular direction. The wasp has velocity. It goes in a straight line to its target. When we have to **specify** (say) in which direction something is going as well as its size, this is called a **vector** quantity.

The fly is going round in circles. Its direction is changing all the time. It has speed.

Look at the racing car going round this circuit.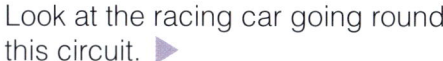

The car travels east from the starting grid. It is travelling due east and has a velocity of 300 km/h. The diagram shows its velocity (in an easterly direction) at different parts of the circuit.

For a complete lap, the direction has changed through a complete circle. The car has a lap speed of 300 km/h.

The velocity in an easterly direction over the whole lap

$= 300 + 0 + -300 + 0 = 0$ km/h

It is the change in direction over the whole lap that makes the lap velocity zero.

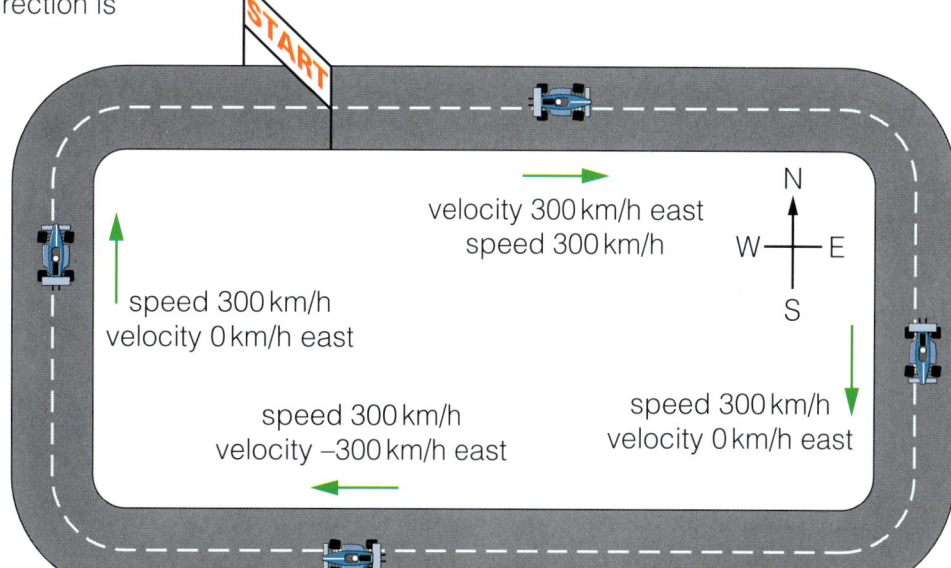

Q1 What is speed?

Q2 What is velocity?

Q3 What do you need to know to work out the speed of a bicycle?

Q4 Why has the wasp got a velocity and the fly cannot be given a velocity?

Q5 A town is 50 km away. Which is more likely to travel towards it at a steady velocity of 80 km/h, a helicopter or a car? Explain why.

4 Balanced forces and velocity

Measuring velocity

Let's build some model cars and measure their velocity. We shall use pulleys and belts to drive the road wheels.

Belt and Pulley Drive

Q1 Copy this table.

Drive system	Time to travel 2 m (s)	Velocity (m/s) = $\frac{distance\ (m)}{time\ (s)}$
Belt and pulley drive		
Compound belt drive		

Apparatus

- [ ] stop watch [ ] metre rule
- [ ] battery and battery holder
- [ ] these parts from LEGO Technic 2 kit 1032 [ ] 2 long wires

A Build the chassis of this belt driven car. ▼

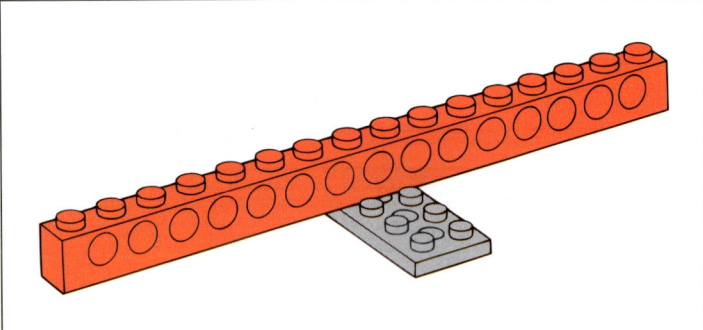

B Add the axles and a pulley. They are eight studs long. Complete the chassis. ▼

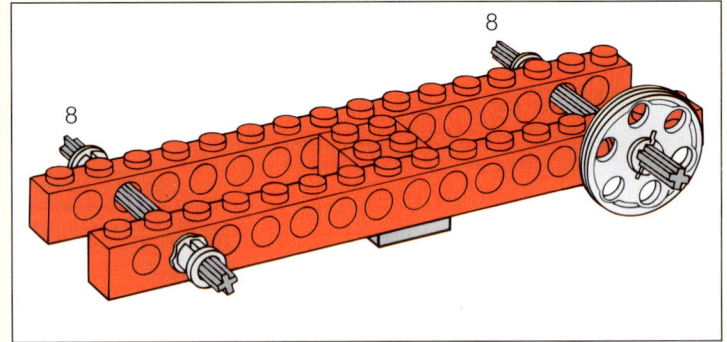

C Add the motor, drive belt, four wheels and tyres. ▼

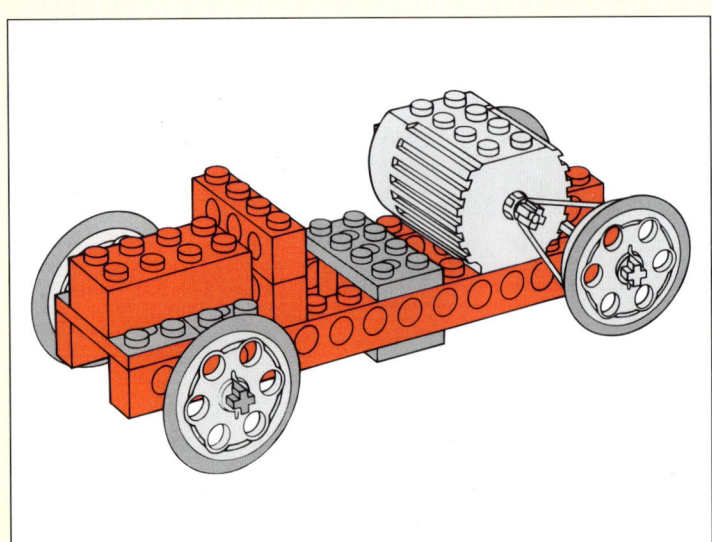

D Connect the motor to the battery unit. ▼

4 Balanced forces and velocity

E Find a clear space about 3 m long. Mark out a test track. The start and finish lines should be 2 m apart. ◀

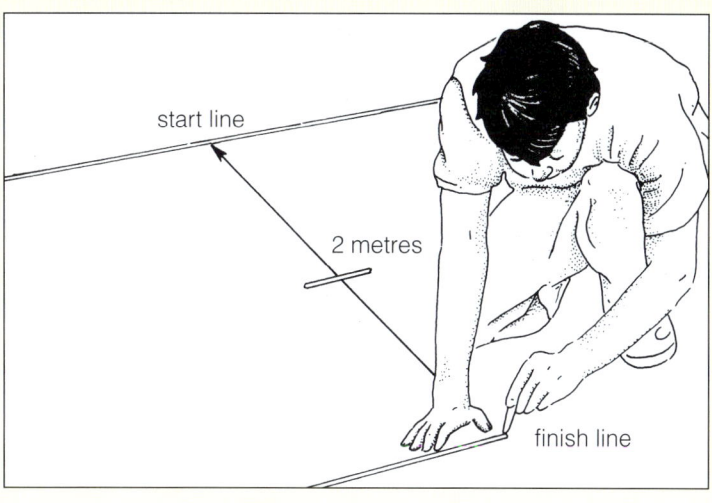

F Put the car about 20 cm behind the start line, and hold the stop watch. Switch on the motor. Start the watch when the front wheels cross the start line. Stop the watch when the front wheels cross the finish line. Record your results in the table. Work out the velocity. Repeat **F** twice. ▼

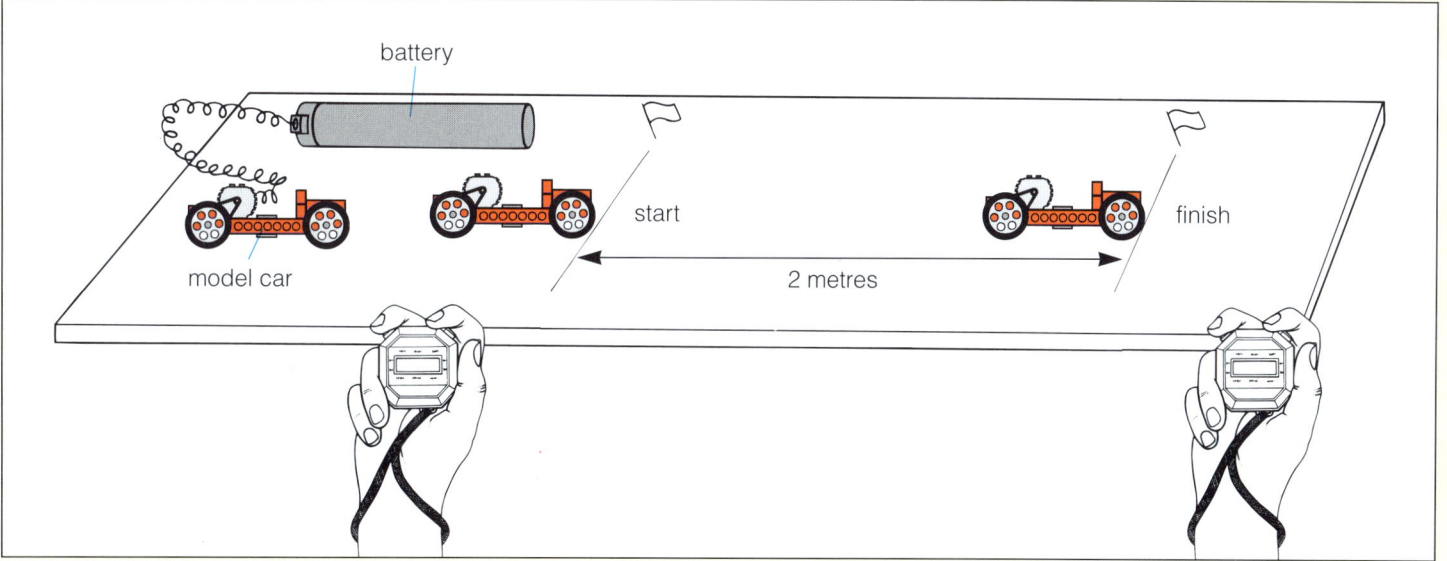

G Change to a compound belt drive as shown. Repeat **F**. ▼

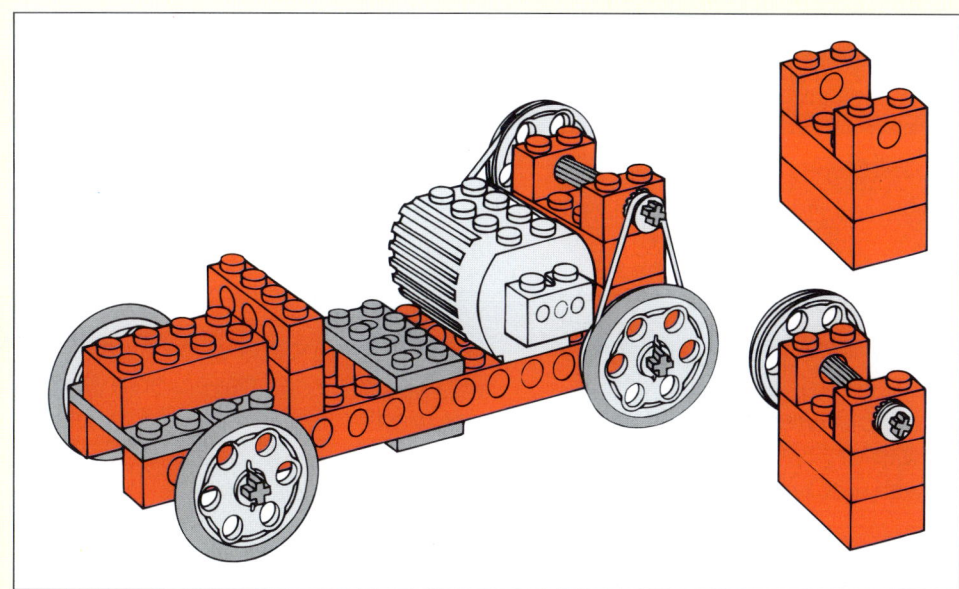

Q2 What was the velocity of the faster car?

Q3 What was the velocity of the slower car?

Q4 What could you do to the test track to make your results more accurate?

Q5 Pat drives north for 3 hours and travels 150 km. The return journey takes 2 hours.
a What is Pat's velocity when travelling north?
b What is the velocity on the homeward journey?
c What is the average speed for the whole journey?

5 Friction forces

Gripping the road

Let's see how good your model is at gripping the road.

Q1 Copy this table.

Car with:	Force to move car (friction) (N)
no tyres	
dry tyres	
wet tyres	

Apparatus

- ☐ your model car from page 10
- ☐ newton meter
- ☐ large tray

A Remove the tyres from the car. Hook the newton meter on to the car. Put the car in the tray. Gently pull until the car just begins to slide. Note the reading on the newton meter. ▼

B Replace the tyres. Repeat **A**. Complete the table.

C Add a little water to the tray. Repeat **A**. Complete the table.

Improving the grip on the road

All racing cars have wide tyres made from a very sticky rubber. The tyres are smooth and do not have a tread on them. The tyre rubber is soft and soon wears away. They are usually changed two or three times in a 250 mile race. With a harder rubber, the grip or friction force would not be as good. ▶

The smooth 'dry' tyres are very dangerous on wet roads. A layer of water builds up under the tyre. It has a low friction force. The car is said to **'aquaplane'**. The pit crew must change the tyres to 'wet' tyres before this happens. These are more like ordinary tyres with a tread pattern on them. The grooves in the tread let the water run away from under the tyre. The tyre makes contact with the road.

▲ Tyres on ordinary cars must be safe on wet and dry roads. They must last for 20 000 or 30 000 miles. All road tyres should have a tread of 1.6 mm on them. A worn bald tyre does not channel the water away. The car will aquaplane. A dangerous skid will soon happen.

Q2 What force was needed to make the car slide:
a without tyres
b with dry tyres
c with wet tyres?

Q3 What happens when a car with smooth tyres travels on a wet road?

Extension exercise 4 can be used now.

6 Unbalanced forces

Apparatus
- [ ] straw [ ] bulldog clip
- [ ] nylon fishing line [ ] 2 G-clamps
- [ ] clamps and stands
- [ ] sausage-shaped balloon

Forced to move

In this experiment you are going to see how an unbalanced force makes a balloon **accelerate** (go faster). The force of the escaping air in one direction makes the balloon move in the other direction.

A Blow up a balloon and seal the end with a **bulldog clip**. Fix the balloon as shown. Tie each end of the nylon line to stands clamped to the bench. Make sure the line is **taut** and that the balloon is free to slide.

B Pinch the neck of the balloon with your fingers and remove the bulldog clip. Let the balloon go.

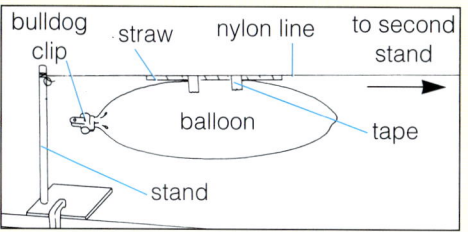

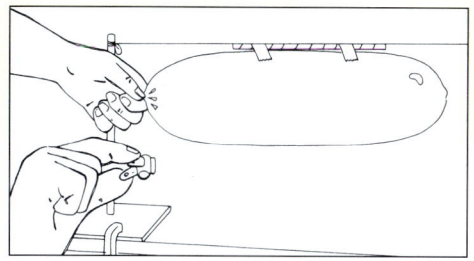

Jet propulsion

There are two types of jet engines: **turbojets** and **turbofans**. Turbojets work like the balloon. They push a jet of hot air out of the back of the aeroplane. The unbalanced force this produces pushes the plane forward. Turbojets are very noisy and are used in Concorde and fast military aeroplanes

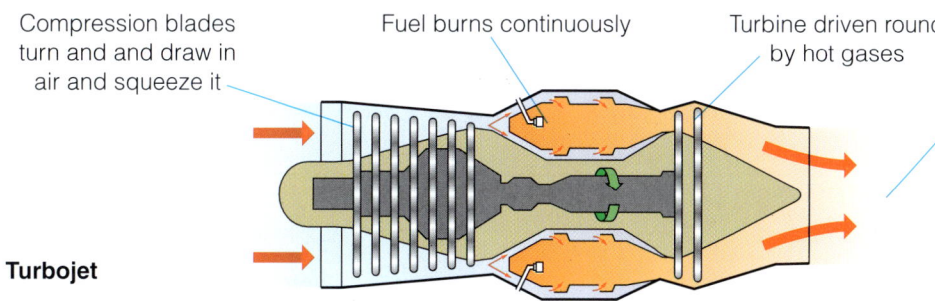

Turbojet

Most aeroplanes use turbofans. These use a big fan to push out a lot more air at a slower speed. They are quieter and more **economical**. Fuel is burnt in the core of the engine. The hot gases drive the turbine and the fan. Some of the air from the fan goes into the engine to burn the fuel, the rest bypasses the engine and is blown out of the rear of the aeroplane. This bypass air provides most of the thrust at low speeds.

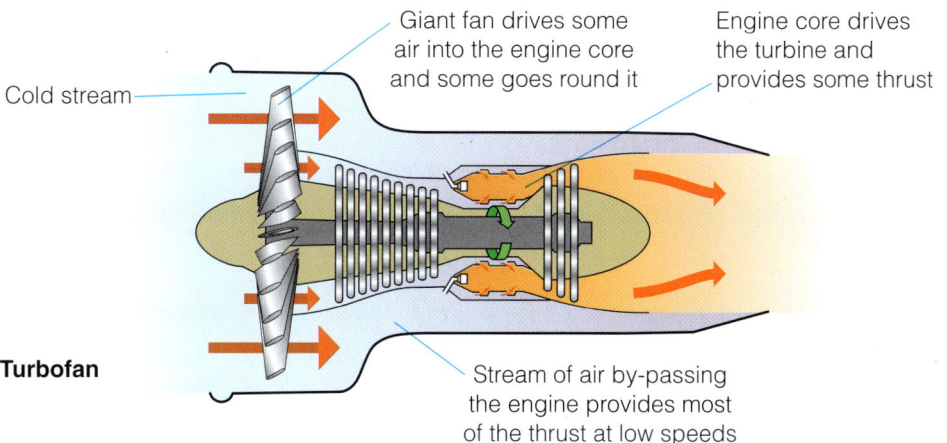

Turbofan

Q1 What happened when you removed the bulldog clip and let the balloon go?

Q2 Draw two copies of the balloon. On one, show the balanced forces acting on the inside of the balloon before release. On the other, show the forces acting when it is released.

Q3 Does an unbalanced force act on the balloon? Give a reason for your answer.

Q4 Why do airlines use turbofans rather than turbojets?

6 Unbalanced forces

Forced to move in a circle

Let us see what happens when we fix your model car (from page 10) to a central pivot.

A Use the nylon line to fasten your model car to the clamp stand. Switch on. Note how the car moves round the stand. ▼

Apparatus
- [ ] your model car with battery and connecting wires
- [ ] clamp stand
- [ ] nylon line

A force in the string pulls the car round in a circle. This force, which acts towards the centre of the circle, is called the **centripetal force**.

If the string breaks there is no centripetal force. The car moves in a straight line. Your teacher might demonstrate this by using a match to burn the nylon line.

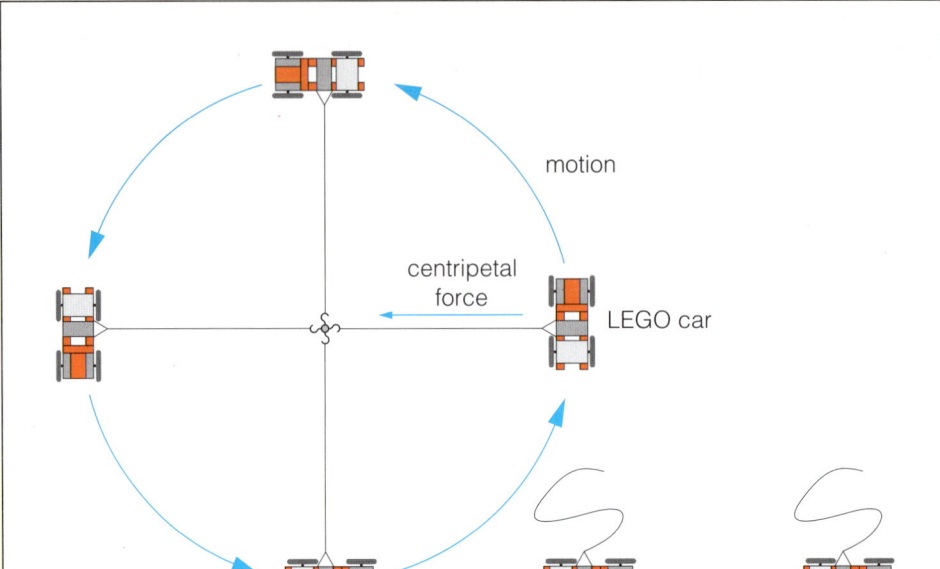

Here are some other examples of objects moving in a circle. ▼ ▶

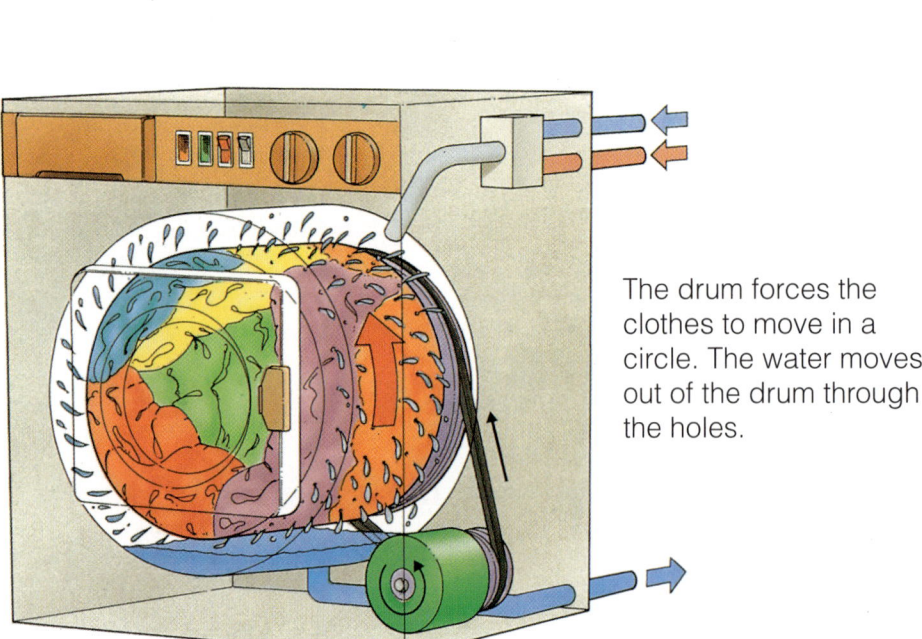

The drum forces the clothes to move in a circle. The water moves out of the drum through the holes.

Q1 The string provides the centripetal force to make the car move in a circle.
a What provides the centripetal force in each of the pictures above?
b In which direction does the centripetal force act?

Q2 The chair-o-planes in the photograph turn at a steady speed. What happens to their velocity?

14

7 Unbalanced forces and acceleration

Unbalanced forces change velocity

When you first get on a bicycle your velocity is zero. Press hard on the pedals and your velocity increases. As you go faster and faster, we say you are **accelerating**. This acceleration has been caused by an unbalanced force **transmitted** from your leg muscles, through the pedals and chain drive to the tyres on the road.

When you put your brakes on, you slow down. We say you are **decelerating**. It is the opposite of accelerating. Sometimes we say it is a negative (−) acceleration. The deceleration is caused by an unbalanced force transmitted from your finger muscles to the brake mechanism and to the tyres and the road.

We don't all take the same time to increase our velocity. One of these cyclists would take longer to get to 40 km/h than the other. Their accelerations are different. We say that:

$$\text{acceleration} = \frac{\text{change in velocity}}{\text{time taken for it to change}}$$

at start velocity = u m/s t seconds later velocity = v m/s

Let's suppose the velocity was u m/s at the start, then it increased to v m/s after t seconds.

> The velocity changed from u to v so the change in velocity is $(v − u)$
> Acceleration (a) (m/s^2) = $(v − u)/t$ (m/s)/(s)
> $a = (v − u)/t$

Let's put some numbers in. Suppose the cyclist went from 1 m/s to 10 m/s in 3 seconds.

at start velocity = 1 m/s   3 seconds later velocity = 10 m/s

What is his acceleration?

> Change in velocity $(v − u)$ is 10 m/s − 1 m/s = 9 m/s
> Acceleration = change in velocity/time
> $a = \frac{9}{3} = 3$ m/s^2

Q1 What happens to an object when it accelerates?

Q2 What causes the acceleration?

Q3 What happens when something decelerates?

Q4 What causes the deceleration?

Q5 Give the equation you would use to calculate acceleration.

Extension exercise 5 can be used now.

7 Unbalanced forces and acceleration

Using a ticker timer

The ticker timer has a vibrating hammer. The paper tape passes under the hammer and carbon paper. The hammer puts 50 dots on the tape every second. Let's use the ticker timer to measure velocity and acceleration.

Apparatus
- long wooden ramp
- ticker timer and paper tape
- low voltage a.c. power pack and connecting wires
- dynamics trolley
- scissors
- blocks of wood
- sticky tape

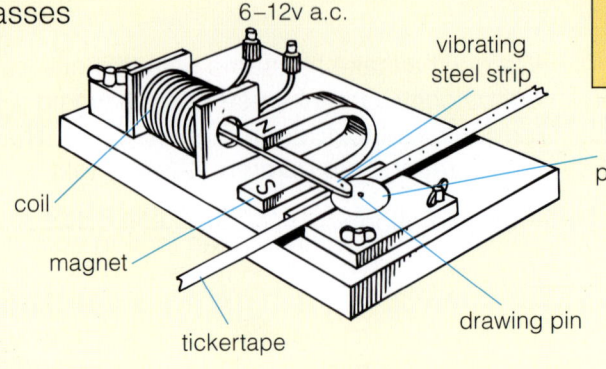

A Use wires to connect the timer to the 6 V a.c. sockets of the power pack. ▼

B Switch on. Pull the tape at a constant velocity. ▼

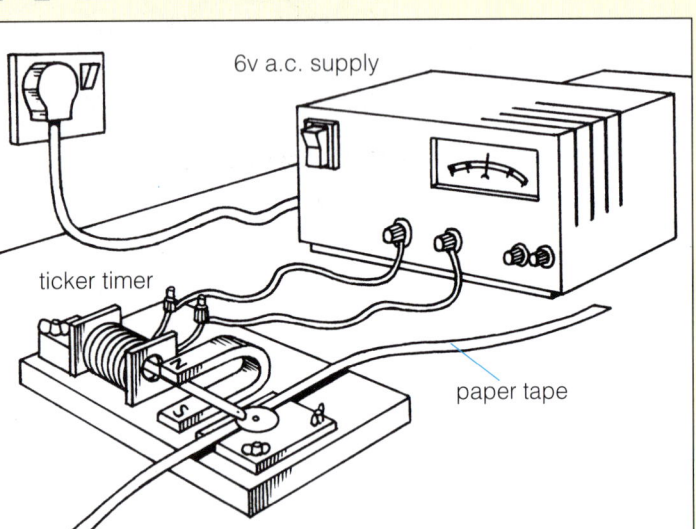

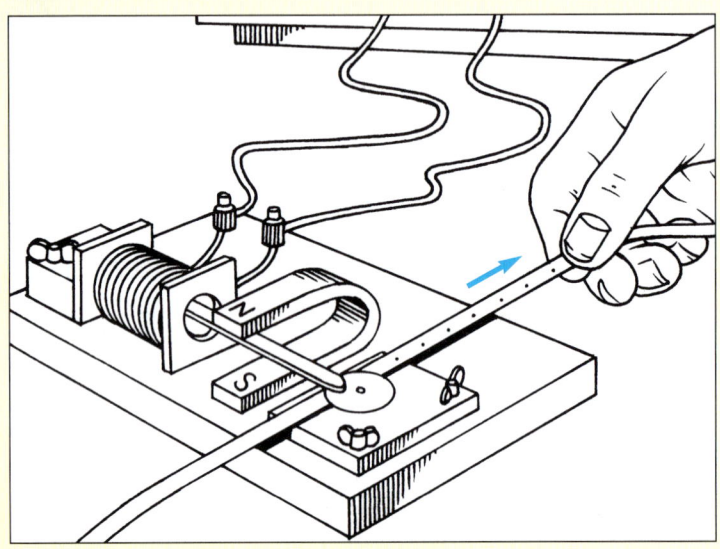

C Look at the marks made on the tape. They should be evenly spaced. Mark every fifth dot. The timer made 50 dots every second so five dots is 0.1 second. ▼

D Get a pair of scissors, use them to cut the tape into five-dot lengths. Keep the lengths in order. ▼

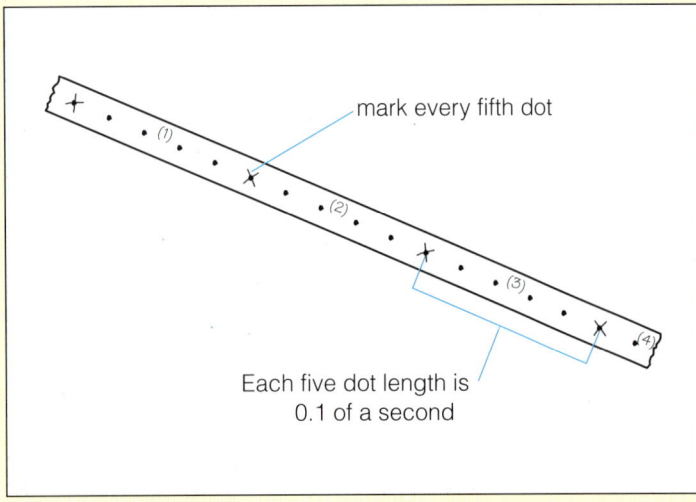

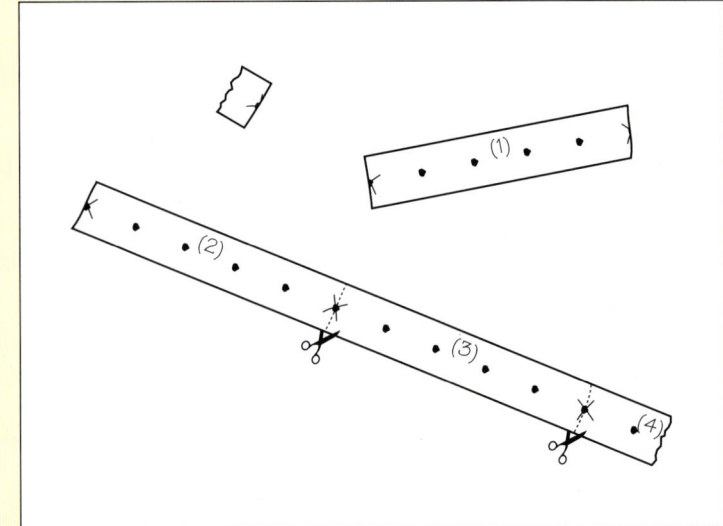

7 Unbalanced forces and acceleration

E Stick the strips side by side in your book. Keep them in order. Each strip is the distance covered in 0.1 second. The strip chart is one of velocity against time. If all your strips are the same length, you pulled at a constant velocity. ▶

F Put blocks of wood under one end of the ramp to raise it to about 150 mm. Use sticky tape to fasten the tape to the trolley. Put the trolley at the top of the ramp. Switch on and release the trolley. ▼

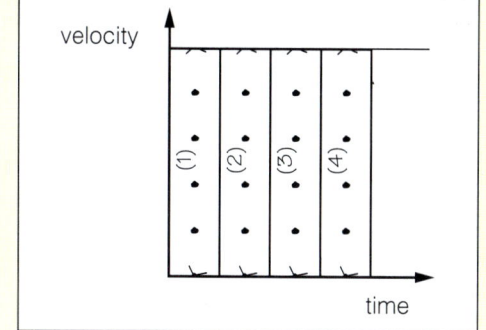

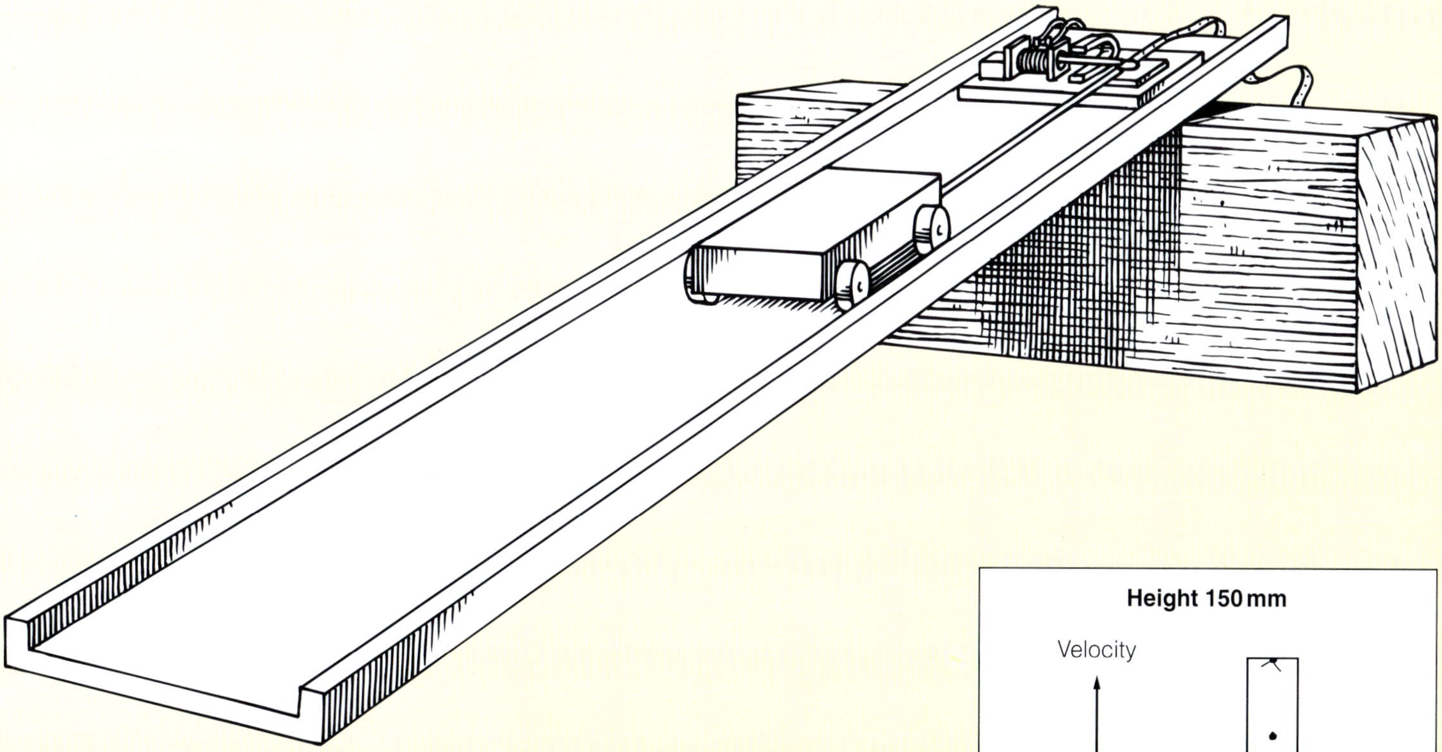

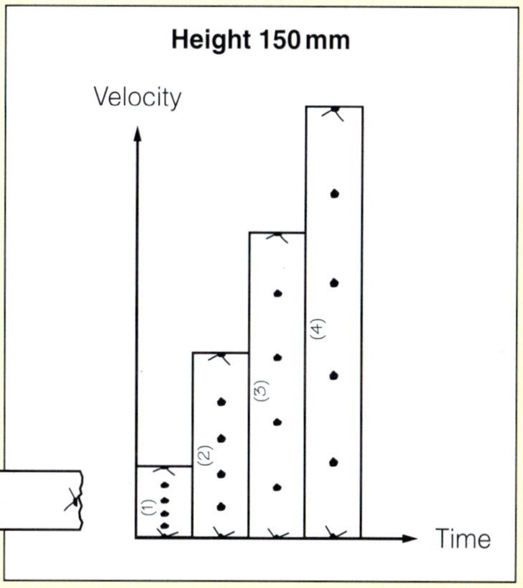

G Repeat **C**, **D** and **E**.

H Raise the ramp to 200 mm. Release the trolley. Repeat **C**, **D** and **E**.

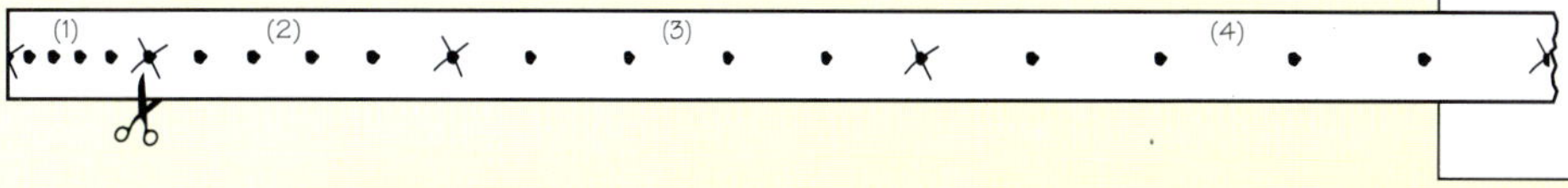

Q1 What is velocity and how is it different from speed?

Q2 How many dots did your timer make in one second?

Q3 How did the slope of the velocity time graph in **H** compare with the graph from **G**?

Q4 What happened to the acceleration when you increased the slope of the ramp?

Q5 What do you think would happen to the spacing of the dots if the trolley began to slow down?

17

7 Unbalanced forces and acceleration

Measuring the velocity and the acceleration

Apparatus
- strip charts from your experiment
- rule

Velocity

Let's find out what the velocity of the paper was when you pulled it.

A Measure the length of the five-dot strip. This is the distance the tape travels in 0.1 second. Let's suppose it is 5.6 cm. ▶

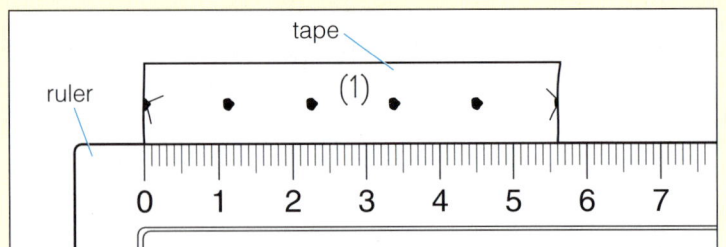

$$\text{Velocity (cm/s)} = \frac{\text{distance travelled (cm)}}{\text{time taken (s)}}$$

$$\text{Velocity} = \frac{5.6 \text{ (cm)}}{0.1 \text{ (s)}} = 56 \text{ cm/s}$$

Now do the same for your tape.

Acceleration

Choose two of the longer strips.

Measure the length of one strip. Measure the length of the strip next to it.

Let's suppose the length of the first strip (A) is 5.2 cm long and the one next to it (B) is 6.4 cm long.

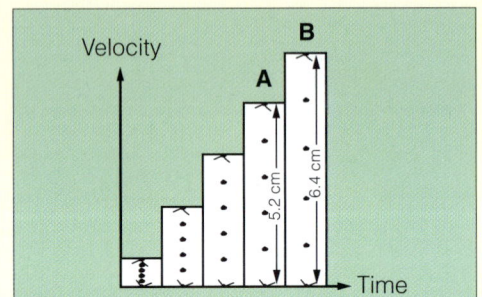

Find the velocity of the first or initial strip.

For A $\text{Initial velocity} = \dfrac{\text{distance (cm)}}{\text{time (s)}}$

$\text{Initial velocity} = \dfrac{5.2 \text{ (cm)}}{0.1 \text{ (s)}} = 52 \text{ cm/s}$

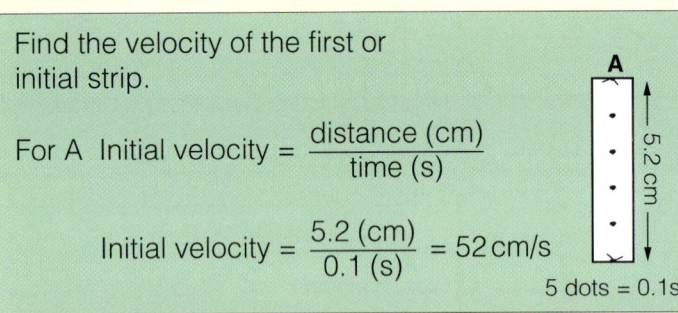

5 dots = 0.1s

Now find the velocity (final) of the next strip.

For B $\text{Final velocity} = \dfrac{\text{distance (cm)}}{\text{time (s)}}$

$\text{Final velocity} = \dfrac{6.4 \text{ (cm)}}{0.1 \text{ (s)}} = 64 \text{ cm/s}$

5 dots = 0.1s

Acceleration

Remember that acceleration = change in velocity/time

Now the velocity changed from 52 cm/s to 64 cm/s.

The change in velocity is 64 − 52 = 12 cm/s

One strip is 0.1 second from the next

$$\text{Acceleration} = \frac{12 \text{ (cm/s)}}{0.1 \text{ seconds}} = 120 \text{ cm/s}^2$$

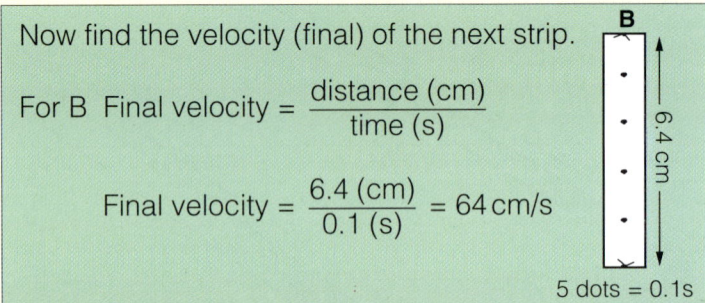

initial — velocity = 52 cm/s

final (0.1 seconds later) — velocity = 64 cm/s

Now do the same for your strips. It is more accurate to take the fifth strip from the initial velocity as the final velocity. Now the time taken to accelerate will be 5 × 0.1 or 0.5 seconds.

Q1 What was the constant velocity of your tape in the first experiment on page 17?

Q2 In the second experiment on page 17, what was the acceleration?

Q3 In the third experiment when you increased the slope, what was the acceleration?

Extension exercise 6 can be used now.

8 Forces on falling objects

A paper helicopter

Two forces act on a falling body. The weight force acts downwards and air resistance acts upwards. You are going to make a paper helicopter and find out how increasing the downward force on it and reducing air resistance changes its speed.

Apparatus

☐ piece of thin card ☐ ruler
☐ scissors ☐ paperclips ☐ pen
☐ stop watch

Q1 Copy this table.

Number of paper clips	Time of fall (seconds)

A Draw this shape. Cut it out along the solid lines. The dotted lines are fold lines. ▼

B Fold along the dotted lines to make the helicopter. Put a paperclip on the bottom. ▼

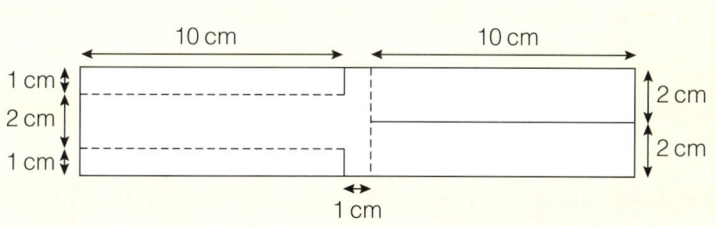

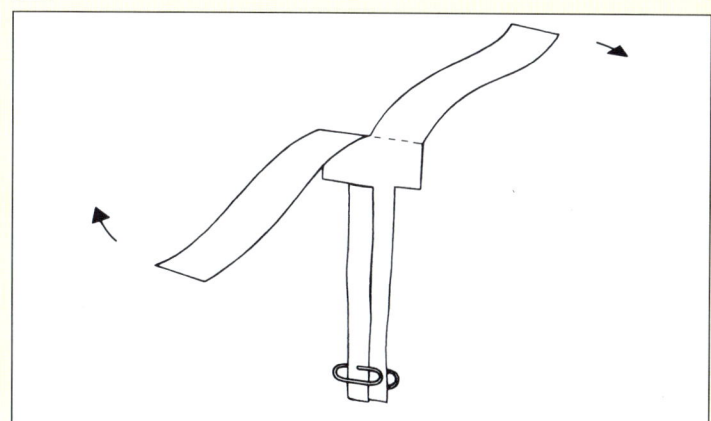

C Standing on a stable surface hold the model 2 metres above the floor. Measure how long it takes to reach the floor. Add a second paperclip and repeat **C**. Add a third paperclip and repeat **C**. Complete the table. ▼

D Now shorten the wings. Watch what happens when you drop it this time. ▼

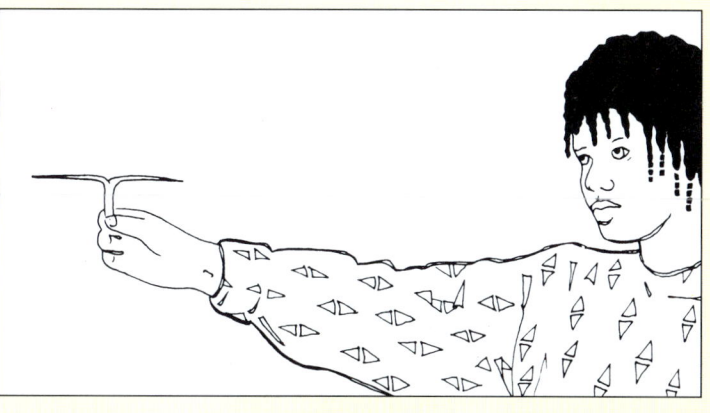

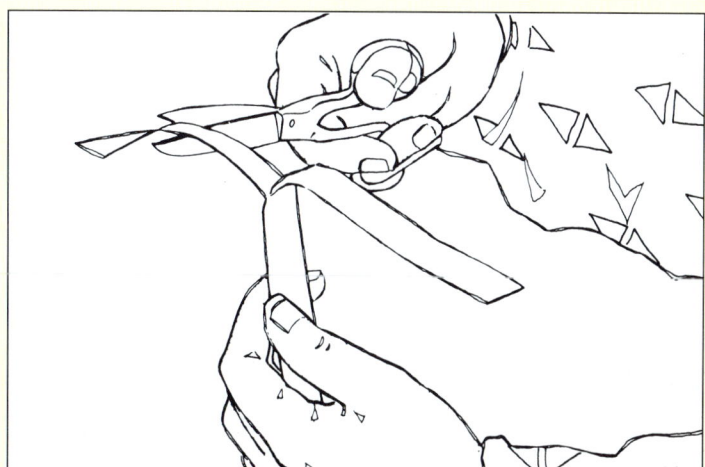

Q2 How did you increase the downward force?

Q3 When you increased the downward force did the speed increase, decrease, or stay the same?

Q4 What was the effect of making the wings smaller in **D**?

8 Forces on falling objects

Falling slowly

Two forces act on the paper helicopter. Its weight acts downwards and **air resistance** acts upwards. At first the air resistance is small. The weight force is unbalanced. The paper helicopter accelerates. As it speeds up, the air resistance force increases and the acceleration is not as great. Soon the two forces balance and the helicopter falls to the ground with a constant (**terminal**) velocity. Let's investigate parachutes.

Apparatus

(You may need some or all of these. You may ask for any other apparatus you need.)
- ☐ polythene bin bag
- ☐ tissue paper ☐ newspaper
- ☐ scissors ☐ felt pen
- ☐ cotton thread ☐ paperclips
- ☐ stopwatch ☐ small weights
- ☐ rule ☐ sticky tape ☐ labels

Investigation: Safe parachutes

Your firm wants to make a range of parachutes to drop goods of different weights safely. You need to find out what things make a safe drop.

Sorting out the problem

In your group decide:

- which things affect the rate of fall of your parachute, for example: shape, size;
- how you are going to change them and make it a fair test;
- what is a safe speed: too fast and the goods might break; too slow and they might drift away on the wind;
- how you can make the parachute stable so that it doesn't sway from side to side.

Planning the investigation

Think about the measurements you will take and how you will record them. Discuss your ideas as a group. Write your ideas in your book.

Carrying out your investigation

Make your parachutes. Test them in a scientific way. Record your results.

Evaluating your findings

Report your findings and suggest improvements to your designs.

8 Forces on falling objects

Air resistance

All objects are pulled towards the centre of the Earth. As an object falls it goes faster (accelerates). Soon the air resistance limits the **acceleration**. It reaches a constant speed called the **terminal velocity** when these two forces balance.

▶ A **skydiver** falls freely from an aeroplane. The skydiver's weight acts downwards and air resistance pushes upwards. This slows her fall. The terminal velocity is about 50 m/s (190 km/hour).

▶ At a safe height the skydiver pulls the rip cord and the parachute opens. The parachute has a large air resistance. The terminal velocity is about 8 m/s or less. The skydiver lands safely. The bigger the area of the falling body, the greater the air resistance and the slower the terminal velocity.

If there is no air there will not be any air resistance. If there is no air resistance an object will fall at an ever increasing speed. It will not reach a steady terminal velocity. This increase in speed is called the 'acceleration due to gravity'. On Earth the speed increases by about 10 m/s every second. More accurately measured as 9.81 m/s^2.

▶ Astronaut David Scott drops a hammer and a feather from the same height above the Moon's surface. Both hit the surface at the same time.

Q1 Explain what is meant by air resistance.

Q2 How can you make an object fall through the air more quickly?

Q3 Skydivers jump from the aeroplane at different times. They can meet up with each other during their fall. Explain how they can do this.

Q4 Why do the hammer and feather fall at the same rate on the Moon but not on Earth?

Extension exercise 7 can be used now.

9 Pressure

Under pressure

Lying on a mattress is very comfortable, a bed of nails is not. On the mattress your weight is spread over a large area. On the nails it is painfully concentrated on a few sharp points of very small area and the pressure is high. When a force acts on a surface the spread of the force is called **pressure**.

The unit of pressure is called the pascal (Pa). It measures the force (N) acting on a square metre.

We can calculate it from:

$$\text{pressure (N/m}^2\text{)} = \frac{\text{force (N)}}{\text{area in contact with the force (m}^2\text{)}}$$

Our atmosphere exerts a pressure on us. It is caused by the weight of air above us. At sea level it is about 100 000 N/m². About the same force that a 10 tonne mass would exert on a small bedroom window! Fortunately, the air pressure acts on both sides of the window.

As you go up a mountain there is less air above you. The pressure is much lower because you have a smaller column of air above you. At 3500 m you need five times as much air to give you the same amount of oxygen that you got at sea level. Mountaineers need to let their bodies **acclimatise** (get used to) the altitude.

Let's see what happens when we remove the air from inside a can. The pressure will not be balanced. Your teacher may demonstrate this experiment.

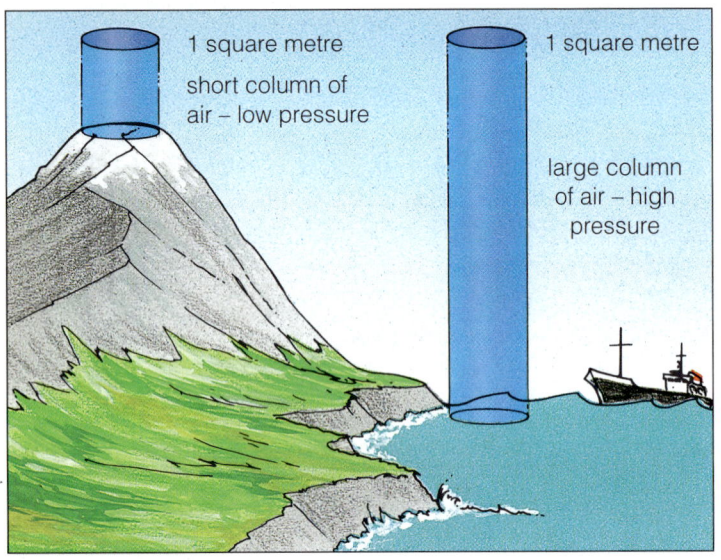

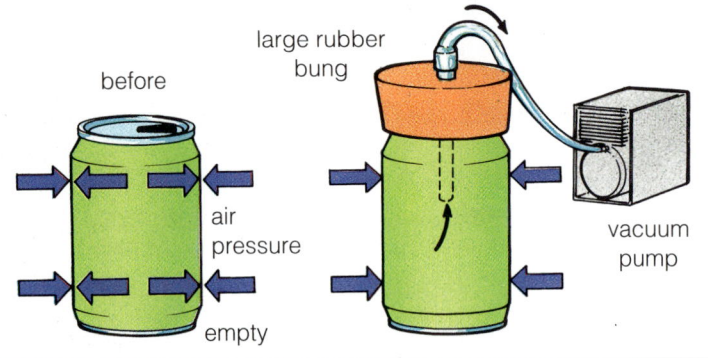

The vacuum pump removes air from the inside of the can. The pressure is not balanced. Atmospheric pressure crushes the can.

Q1 What is pressure? How is it related to force?

Q2 What causes atmospheric pressure?

Q3 Why is the atmospheric pressure less at the top of a mountain?

Q4 If you have an aneroid barometer, measure the atmospheric pressure at ground level and again at the top of your school. Record the pressure readings in both places. What is the pressure difference?

9 Pressure

Pressure from the air

Let's try some simple experiments to show the effects of air pressure

Apparatus
- ☐ a large newspaper
- ☐ large stick/brush handle
- ☐ bowl of water ☐ plastic beaker
- ☐ empty syrup tin with a hole in the lid and in the can

A Put the newspaper over the stick. Knock the free end of the stick down. Feel the air pressure resisting the upward movement of the newspaper and stick. ▼

B Fill the beaker underwater in the bowl. Raise the beaker up and look at the water levels as you do this. ▶

C Fill the tin can from the bowl. Replace the lid. Lift the can out of the water. You can control the drips by gently lifting and lowering your finger from the hole in the lid. ▶

The obedient drip

Q1 What did you feel when you pressed hard on the stick in **A**? Explain why this was so.

Q2 What happened to the water levels as you raised the beaker in **B**? Explain why this happened.

Q3 How did you control the drips in **C**? Explain why this happened.

Q4 Look at the diagram.

Explain how air pressure allows you to drink through a straw.

9 Pressure

Pressure in liquids

You may have noticed that the pressure on your ears increases when you dive underwater in the swimming pool.

Let's find out what pressure depends on. Your teacher may demonstrate these experiments.

Pressure and depth

The tall can has three equal sized holes in it. The can is filled with water.

Q1 Where does the water come out fastest?

Q2 Where is the pressure greatest?

Q3 What happens to the pressure as you increase the depth of the water?

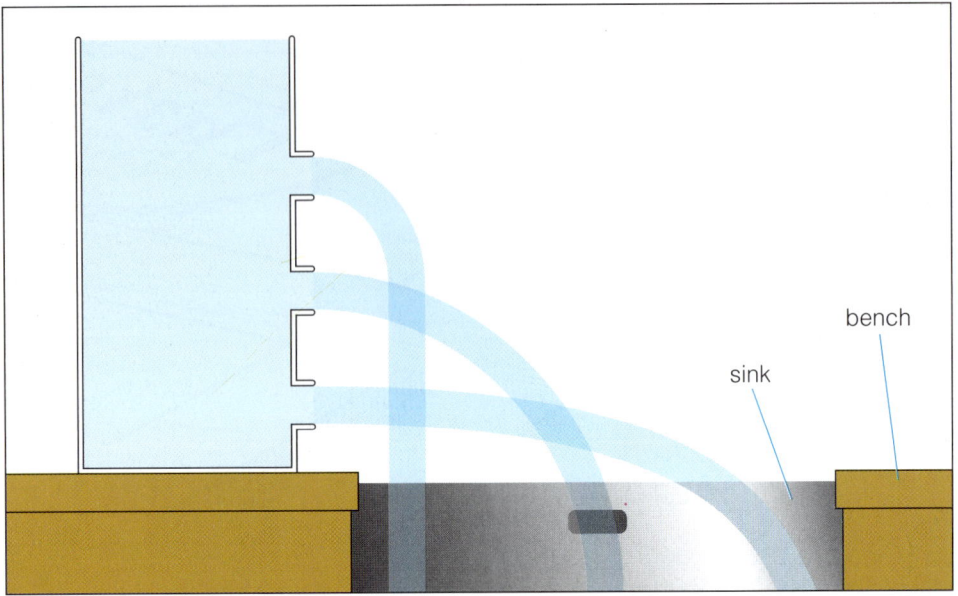

Which direction does the pressure act in?

Your teacher will fill a polythene bag with water, then use a pin to make some holes in it.

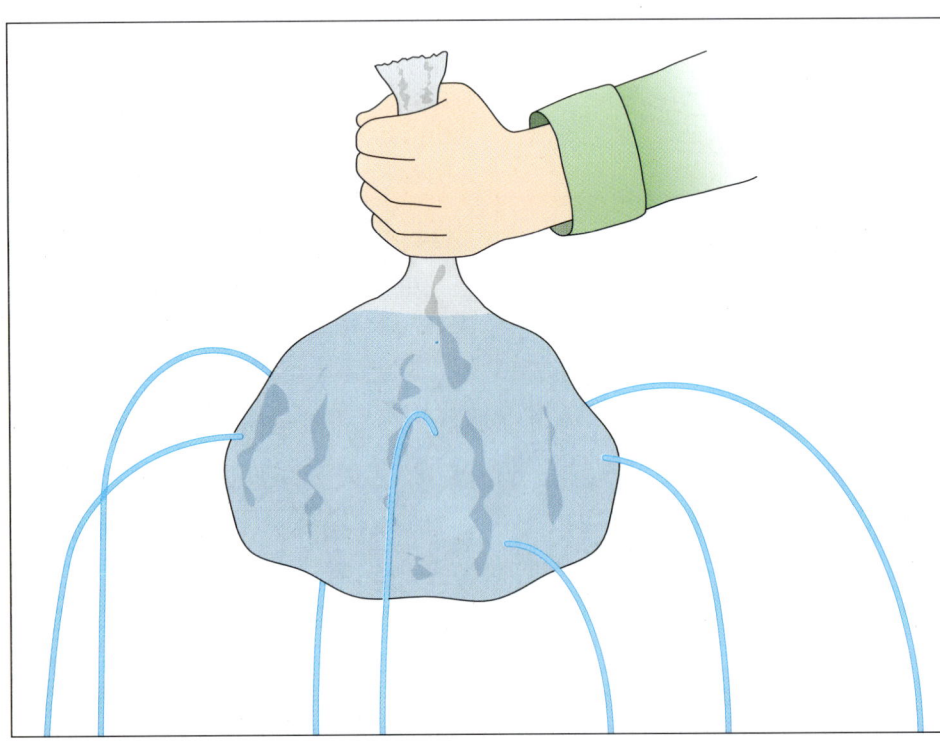

Q4 What happened to the water when the bag was squeezed?

Q5 Which holes did the water come out of?

Q6 Which direction does the pressure act in?

Summary

From these experiments we can say:

- pressure increases with depth; pressure acts in all directions.

- pressure also depends on the density of the liquid and the pull of gravity.

Extension exercise 8 can be used now.

9 Pressure

Applications of pressure

Engineers often build dams to store drinking water or water for hydroelectric power schemes. Pressure increases with depth and acts in all directions. The bottom of the dam gets most pressure. The deeper the water, the thicker the dam must be. It must be strongest at the bottom.

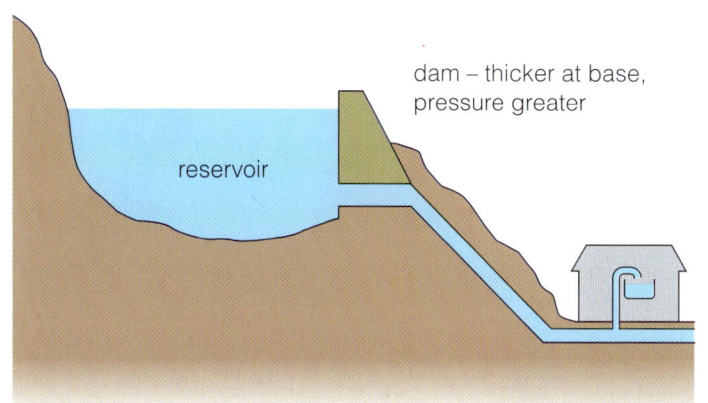

The divers have to breathe compressed air. The deeper they go the greater the pressure from the water. For very deep dives their diving suits must be strong enough to protect them from the great pressure. To explore the deep ocean they have to use a bathysphere or submersible craft capable of withstanding extreme pressures.

Another problem for the divers is that their air supply must come to them at greater pressures. At these pressures more nitrogen from the air dissolves in their blood. If the diver comes back to the surface too quickly the nitrogen comes out of the blood as bubbles. This is like bubbles coming out of a fizzy drink when you take the cap off. This is called the 'bends'. It is very painful and can even kill the diver.

Q1 Where is the pressure in a reservoir the greatest?

Q2 Which part of the dam must be the strongest?

Q3 Why must a diver return to the surface slowly?

Q4 Why must a deep sea diver have a strong diving suit?

9 Pressure

Hydraulic systems

A liquid can transmit forces from one place to another. Car brakes use this type of **hydraulic** system. Let's find out how car brakes work.

Apparatus
- ☐ 2 syringes of the same size
- ☐ syringe of a different size
- ☐ plastic tube to fit

A Connect two syringes of the same size by means of a plastic tube. ▼

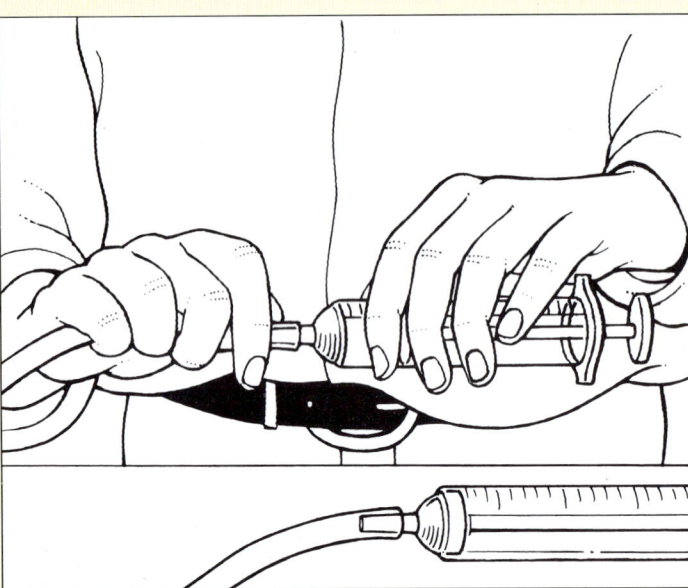

B Remove the pistons and fill the syringes and tube with water. Replace the pistons. ▼

⚠ The floor may become slippery if it gets wet.

C Hold one in each hand and press the pistons with your thumbs. ▼

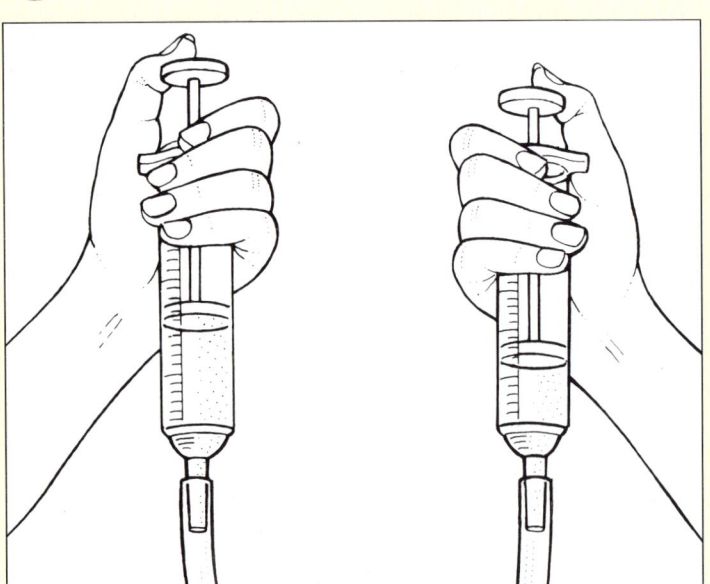

D Change one of the syringes for one of a different size and repeat **B** and **C**. ▼

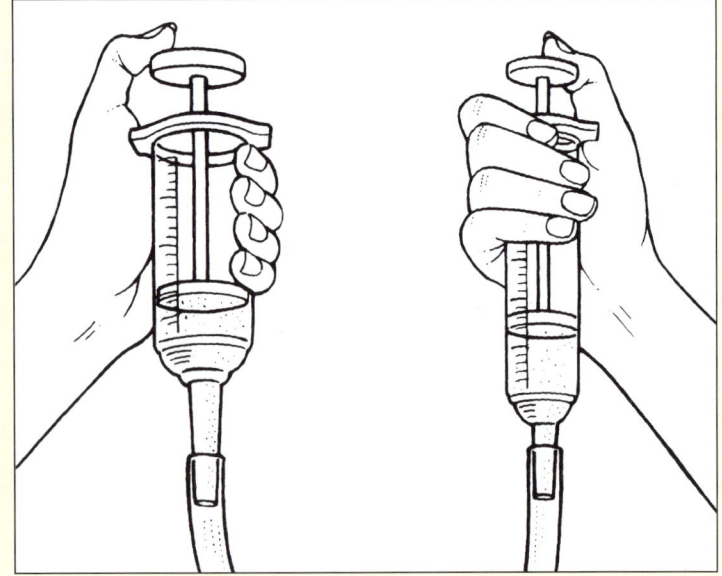

Q1 What difference does it make to the force you feel in **C** and **D** when the syringes are of different sizes?

Q2 Explain the reason for any difference in the forces in **Q1**.

9 Pressure

Hydraulic brake systems

Hydraulic systems are used to transmit pressure or force. The main system operated in this way is the **brake** system in a motor car.

▼ A hydraulic brake system (as shown in the diagram) has a **master** cylinder and **slave** cylinders connected by pipes. The system is filled through the master cylinder with **hydraulic fluid**.

When the **brake pedal** is pushed a piston inside the master cylinder forces the hydraulic fluid into the slave cylinders. The pistons inside each slave cylinder push brake pads onto the brake **drums** or **discs**.

If air gets into the system the brakes will not work very well. Air is **compressible** (can be squeezed into a smaller size) and therefore takes up some of the essential movement in the system.

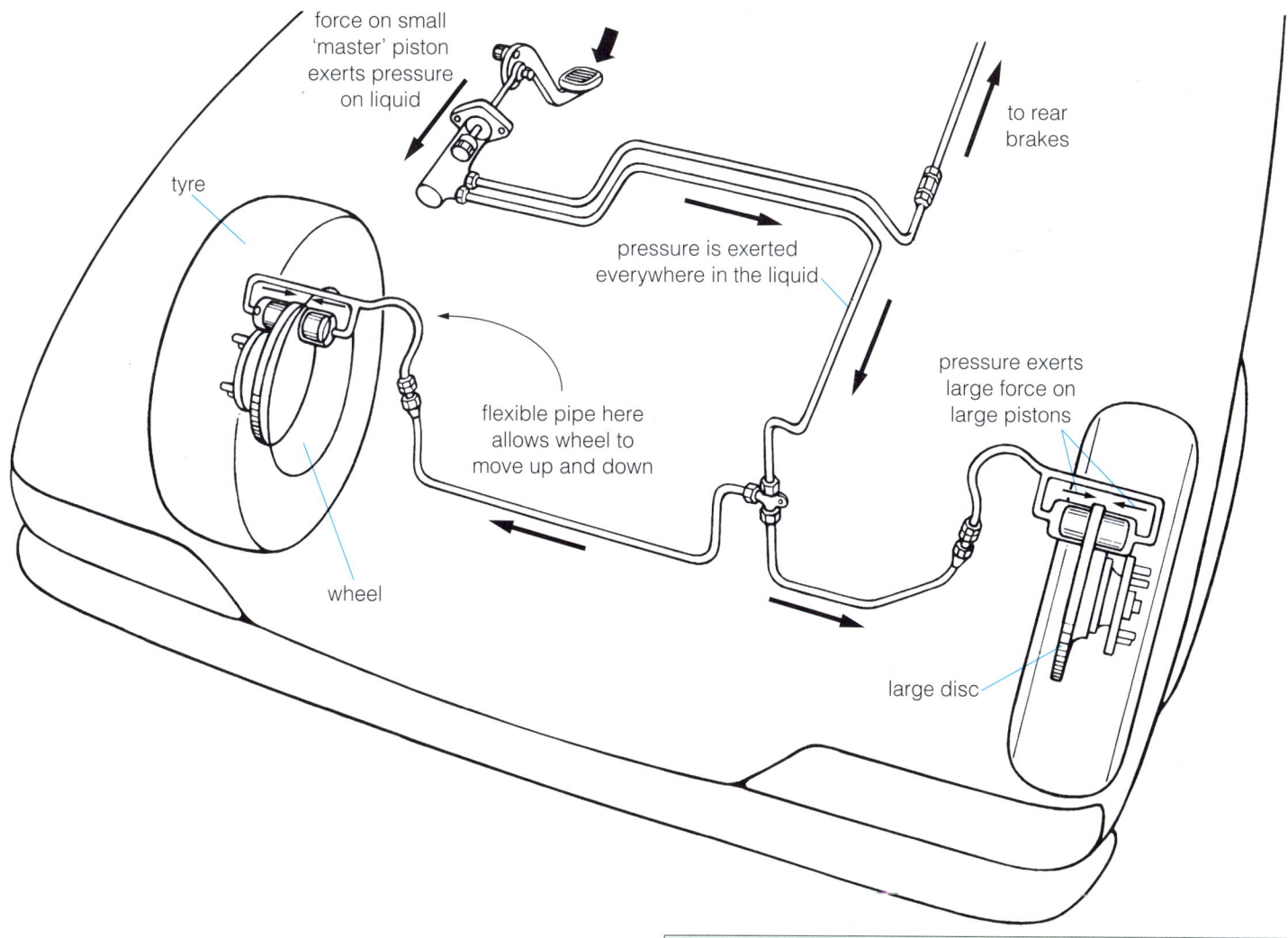

Q1 In a hydraulic braking system one piston is operated by the brake pedal and the other pushes the brakes on. Which one is the smaller piston and which one is the larger piston?

Q2 Why is this?

Extension exercise 9 can be used now.

10 Work, energy and power

Energy

Energy is vital to us all. Something which has energy can do work. Energy is measured in a unit called a **joule**. This is a small unit and we often use kilojoules (1 kJ = 1 000 J) or megajoules (1 MJ = 1 000 000 J).

There are many forms of energy:

Heat or thermal energy	Sound	Movement (kinetic)
Chemical (including food)	Light	Position (potential)
Electric (and magnetic)	Nuclear	Gravitation

We cannot create energy. We cannot destroy energy.
Often we can change energy from one form to another.

This picture shows some energy changes in a fairground ride.

Q1 In what form is the energy supplied to the motor?

Q2 What sort of energy does the car have at the top of the ride?

Q3 What is this converted to as the car rushes to the bottom?

Q4 What forces the car to loop the loop?

Q5 What sort of energy has the car lost at the top of the loop?

Q6 What energy is lost as the brakes are applied? What is the energy converted into?

10 Work, energy and power

Work done and energy

It is hard work carrying bricks up a ladder. Work has to be done. The potential energy of the bricks increases. Let's find out how work and energy are linked.

Let's suppose an apple weighs 1 N. If you pick it up and put it on the table 1 metre high you have done 1 joule of work.

> Work done (joules) = force (newtons) x distance moved (metres) in the direction of the force.

We do one joule of work to lift the apple to the table. Its potential energy has increased by 1 joule.

Sometimes we do not know the weight of an object in newtons but know its mass in kilograms. To convert from mass in kilograms to a weight force in newtons we use:

> force (newtons) = mass (kilograms) × 10 (more accurately 9.81)

(9.81 N/kg is sometimes called the gravitational field strength. It is also a measure of the acceleration due to gravity 9.81 m/s^2.)

Now the apple is about to fall off. It isn't moving yet so its kinetic energy is zero joules but its potential energy is 1 joule.

The apple falls off the table. Its potential energy changes to kinetic energy. Just before it hits the the floor it has lost all its potential energy and has 1 joule of kinetic energy.

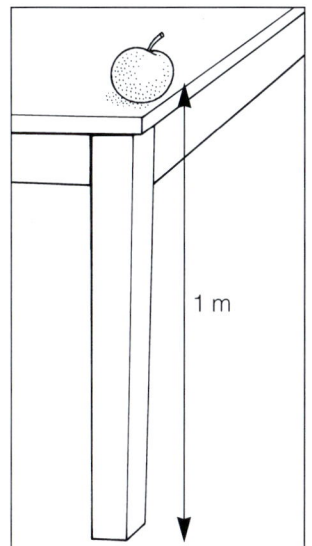
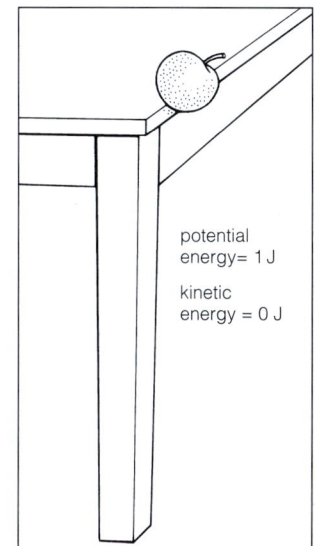
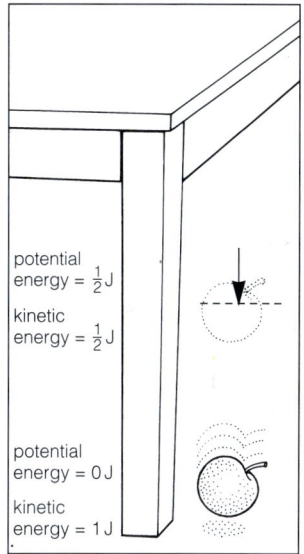

Q1 How much work would be done if two 1 N apples were lifted 2 metres?

Q2 How much work would be done if a force of 5 N was moved 6 metres?

Q3 A person weighs 500 N and stands on a diving board 2 metres above the water. What is her potential energy?

Q4 What is her kinetic energy just before she hits the water?

Q5 What do you think happens to this energy when she hits the water?

Q6 A bricklayer lifts 10 bricks each of mass 2 kg up to a height of 15 m. How much work has been done to lift them?

10 Work, energy and power

Water wheels

We're going to make a water wheel and see how much work it can do. We shall see how powerful it is. A more powerful wheel will do the work in a shorter time.

Apparatus

- ready-drilled cork or rubber bung
- 2 old felt-tip pens
- string
- 8 pieces of plastic for the paddles
- metre rule
- chalk
- small slotted weights and hanger
- stop watch
- 2 clamps and stands
- these LEGO or Meccano parts: axle rod, 2 pulleys

A Push the axle through the cork. ▼

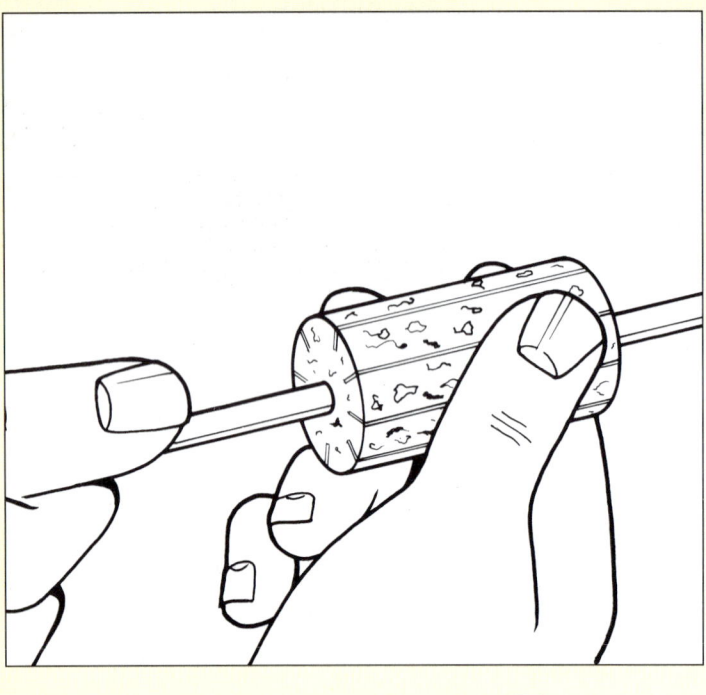

B Fix two pulleys on the axle about 1 cm apart and 1 cm from the cork. Tie a piece of string to one pulley. ▲

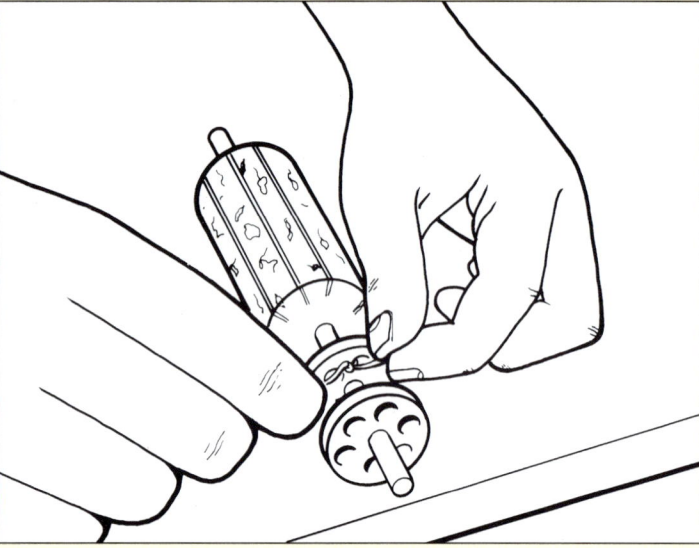

C Fix the plastic pieces in the slots. ▼

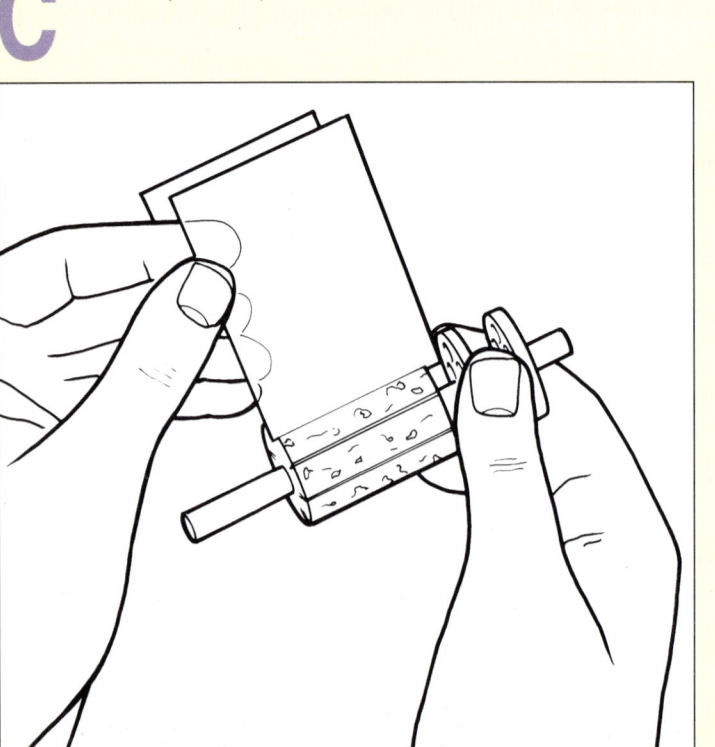

D Use the tubes of the felt-tip pens to hold the wheel between two clamps. Check that the wheel turns freely. ▼

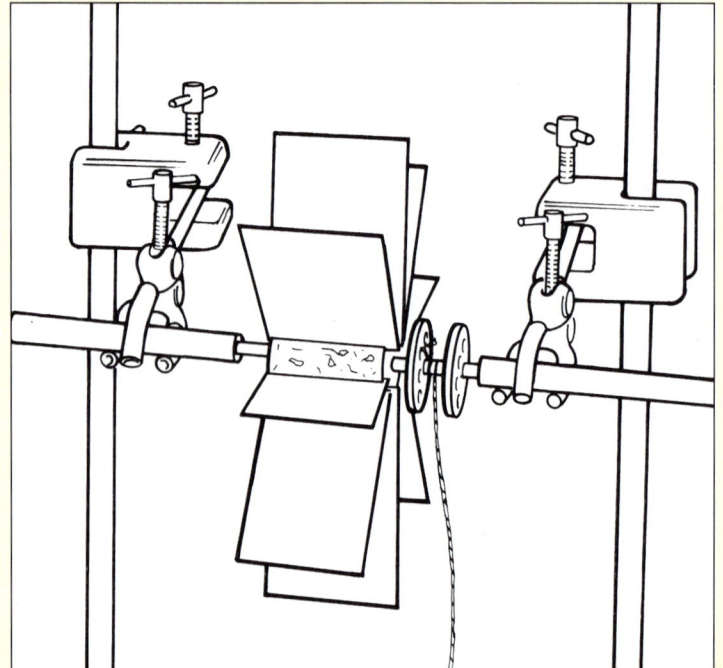

10 Work, energy and power

Testing your water wheel

Q1 Copy this table.

	Trial run No	Weight (N)	Height lifted (m)	Work done (J)	Time taken (s)	Power (W) = $\frac{\text{work done}}{\text{time}}$
Small weight	1					
	2					
	3					
Larger weight	1					
	2					

E Put your water wheel under a tap so that the water will fall on to the paddles. ▼

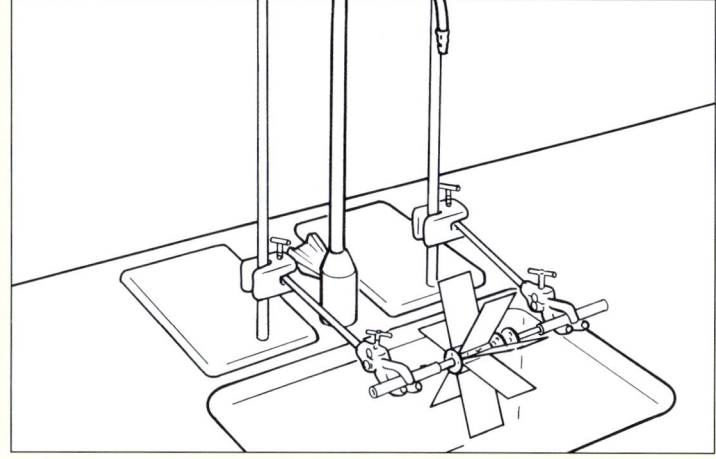

F Check that the string can move freely over the edge of the bench. Fix a small hanger on to the end of the string. ▼

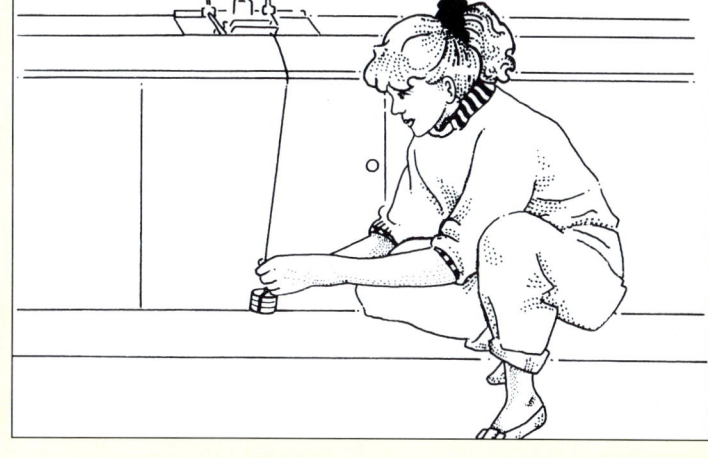

G Use a metre rule and chalk to put a mark on the bench 50 cm from the floor. ▼

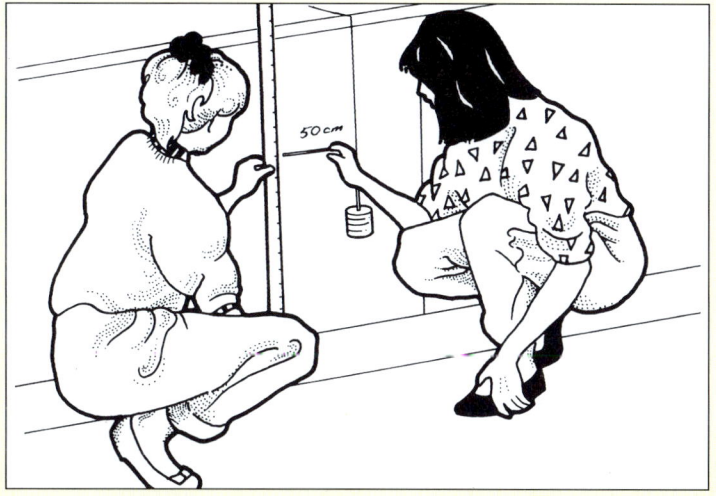

H Gently turn on the water and use your stop watch to find out how many seconds it takes to raise the hanger to the 50 cm mark. Complete the table. Repeat **H** two more times. Increase the weight. Repeat **H** three times. ▼

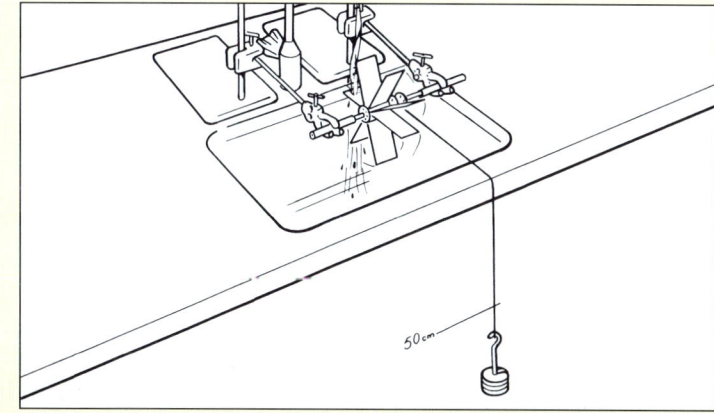

Q2 For the small weight, which trial run was the water wheel:
a most powerful?
b least powerful?

Q3 What was the average time to lift the hanger 50 cm?

Extension exercise 10 can be used now.

10 Work, energy and power

Kinetic energy and power

The water flowing over your waterwheel has kinetic energy. Anything which moves has kinetic energy. If we know its velocity and mass we can calculate the kinetic energy from:

Kinetic energy (joules) = $\frac{1}{2}$ mass (kg) × velocity (m/s)^2 or kinetic energy = $\frac{1}{2}mv^2$

Increasing the speed has a great effect on the kinetic energy.

▼ A car travels at 10 km/hour.

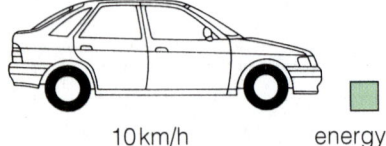

10 km/h   energy

▼ Its speed is doubled to 20 km/h.
Its energy is now $2^2 = 2 \times 2$ or 4 times greater.

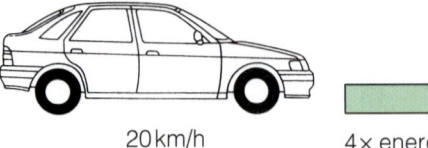

20 km/h   4× energy

▼ If its speed is increased to 30 km/h.
Its energy is now $3^2 = 3 \times 3$ or 9 times greater.

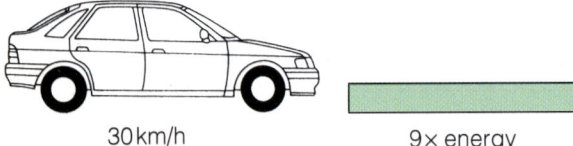

30 km/h   9× energy

▼ At 40 km/h its energy is $4^2 = 4 \times 4$ or 16 times greater. All this energy has to be lost as heat to stop the car. Stopping distances are great at high speeds.

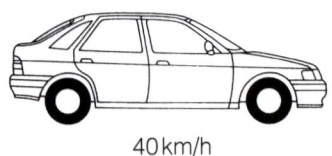

40 km/h   16× energy

▼ This chart shows stopping distances for cars in good condition.

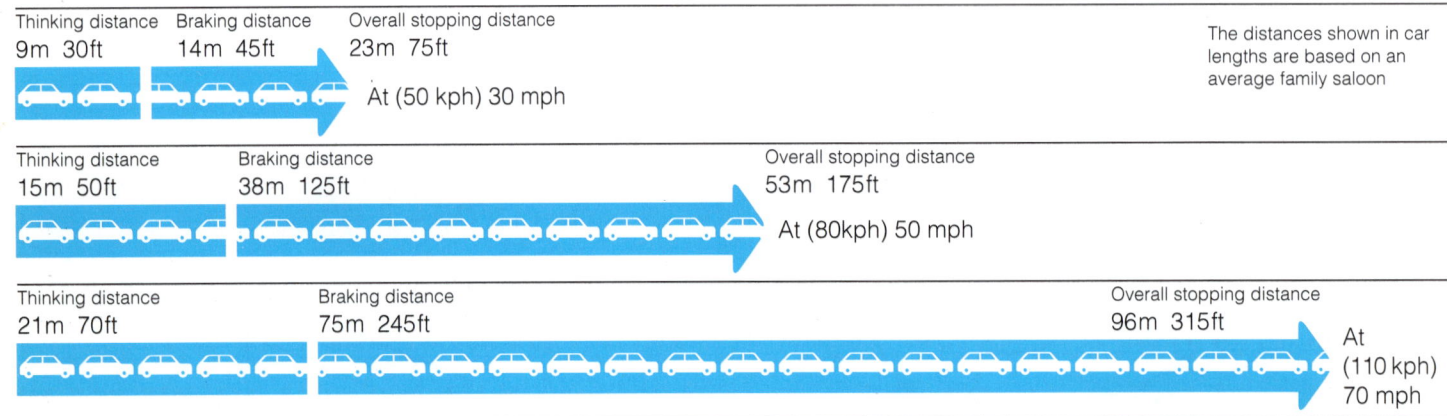

The distances shown in car lengths are based on an average family saloon

Power

This is a word which is often confused with energy or force. In science it has one meaning only. It is the rate at which energy is changed. It is measured in watts.

Power = $\dfrac{\text{energy changed}}{\text{time taken}}$

Powerful brakes change the energy quickly and stop the car in a short time.

Q1 What units are used to measure **a** energy? **b** power?

Q2 What is the difference between energy and power?

Extension exercise 11 can be used now.